Atomic Pair Distribution Function Analysis

Atomic Pair Distribution Function Analysis

A Primer

Simon J. L. Billinge
and
Kirsten M. Ø. Jensen

OXFORD
UNIVERSITY PRESS

Great Clarendon Street, Oxford, OX2 6DP,
United Kingdom

Oxford University Press is a department of the University of Oxford.
It furthers the University's objective of excellence in research, scholarship,
and education by publishing worldwide. Oxford is a registered trade mark of
Oxford University Press in the UK and in certain other countries

© Simon J. L. Billinge and Kirsten M. Ø. Jensen 2023

The moral rights of the authors have been asserted

Published in the United States of America by Oxford University Press
198 Madison Avenue, New York, NY 10016, United States of America

British Library Cataloguing in Publication Data
Data available

Library of Congress Control Number: 2023940074

ISBN 9780198885801

DOI: 10.1093/oso/9780198885801.001.0001

Printed and bound by
CPI Group (UK) Ltd, Croydon, CR0 4YY

Links to third party websites are provided by Oxford in good faith and
for information only. Oxford disclaims any responsibility for the materials
contained in any third party website referenced in this work.

MIX
Paper | Supporting
responsible forestry
FSC
www.fsc.org   FSC® C013604

*This book is dedicated to
all those hardy souls who have
hitherto struggled and become PDF experts.*

The book formerly known as PDF to the people...

Atomic Pair Distribution Function Analysis – A Primer

Simon J. L. Billinge and Kirsten M. Ø. Jensen

with

Soham Banerjee, Emil S. Bozin, Benjamin A. Frandsen,
Maxwell W. Terban and Robert J. Koch

Modern high performance devices, from the latest generation batteries and quantum computers all the way to plastics and pharmaceuticals, utilize materials that are often complex, heterogeneous and whose structure can vary in time (for example, as a battery discharges). Traditional approaches for studying the structure of materials based on crystallography break down in these cases, or only give partial information. In the past 30 years, an alternative approach, atomic pair distribution function, or PDF, analysis has come to the fore as a powerful method for exactly this job. This book is intended for people who think they may want to use PDF methods but are not sure how to get started. It takes you by the hand and leads you through a series of worked examples that go from simple to more complicated, so that at the end of the journey we hope that you will develop the independence and confidence to apply PDF methods to your own research.

In its working form we affectionately titled this book "PDF to the people", a play on the expression "Power to the people" that was popularized in the John Lennon song of that name from 1971. The idea was that PDF was by now somewhat mature, but still difficult for people to adopt and use, and the hope was that this book could bring PDF to the people more easily and empower them, not to mention that the working title also had the same alliteration and similar meter to the original expression. We hope that it does empower people and that it further increases the number of science problems and grad-student PhDs that can benefit from this approach. The authors, at least, will always think of this book as "PDF to the people" regardless of the published title.

The book is intended for graduate students and other research scientists who are new to PDF but want to use it. It is a practical guide and it doesn't aim to teach diffraction theory, structure modeling theory or crystallography. These topics are covered in a very rudimentary way. Rather, it contains a series of examples where the PDF method has been used to yield interesting scientific results, with text describing the steps that were taken to reach that result. It covers many, but by no means all, the main approaches of extracting scientific information from the data. It is based on the philosophy that one learns more by doing than by reading or watching (though these play a role too).

In this spirit, we provide online for download the data and structural models that were used in the examples, so that the reader may try and reproduce (or even improve on) the result that we obtained. The data analysis and modeling requires software which should also be downloaded and installed on your computer. All of the problems can be done using open source and freely downloadable software, though some for-fee software programs that simplify workflow and increase productivity are also illustrated.

We would like to extend a heartfelt thank-you to everyone who has contributed to PDF developments over the years, collaborated with us on scientific projects, whose ideas are the inspiration for the developments. A special thank you goes to Sara Frank for graciously contributing the alternative DIFFPY-CMI solution to the nickel fitting example in the form of a Jupyter notebook. We think readers will find this very helpful. We are also especially grateful to current and former research group members who work so hard and contribute so much without (ok, without much) complaining. They willingly work through the night at synchrotron beamtimes to collect the data and toil tirelessly to analyze, reanalyze, model and remodel the data until we can be certain that a scientific hunch is a reality. Their enthusiasm and passion is what keeps us getting up in the morning. We would also like to thank our families for the patience and fortitude to put up with our long hours and (since Covid) our endless zoom calls. We really appreciate everything you do for us.

Contents

List of figures

List of tables

1

Introduction and overview

Simon J. L. Billinge and Kirsten M. Ø. Jensen

1.1   What this book is not

This book does not describe the theory behind the PDF (pair distribution function) method. Readers are pointed to detailed (Egami and Billinge, 2012; Billinge, 2019), and less detailed (Billinge, 2008a; Billinge, 2013), descriptions of the theory, as well as many texts from which these were taken going all the way back to the Debye Equation (Debye, 1915). It is not a review of applications of the method to nanomaterials (Egami and Billinge, 2012; Keen and Goodwin, 2015; Young and Goodwin, 2011; Billinge, 2008b; Billinge and Kanatzidis, 2004; Lindahl Christiansen *et al.*, 2020a), amorphous materials (Wagner, 1978; Benmore, 2012) or liquids (Furukawa, 1962; Barnes *et al.*, 2003). It is not a comprehensive handbook of different approaches for analysing (Jeong *et al.*, 2001; Qiu *et al.*, 2004; Juhás *et al.*, 2013; Peterson *et al.*, 2000; Soper, 2012; Petkov, 1989) and modelling (Billinge, 1998; Proffen and Billinge, 1999; Farrow *et al.*, 2007; Juhás *et al.*, 2015; Neder and Proffen, 2008; Mcgreevy *et al.*, 1990; Gereben *et al.*, 2007; Tucker *et al.*, 2007; Cervellino *et al.*, 2010; Gagin *et al.*, 2014; Bachir, 2016) data, and it does not present a history of the method (Egami and Billinge, 2012), detailed or otherwise. This book does not deal with procedures for extracting information from diffuse scattering in single crystals, for which we direct readers to the excellent books by Welberry 2010, Nield and Keen 2001, Krivoglaz 2012, and Neder and Proffen 2008. It also does not deal with methods for obtaining and extracting information from 3D-Δ PDF experiments which are described, for example, in Weber and Simonov 2012.

1.2   What this book is

The sole aim of the book is to help new users take their first steps in PDFland by presenting a set of worked PDF problems. The approach is to take a successful scientific study using PDF and provide the data and the steps that were taken to get the result. You are encouraged to try the steps yourself to try and reproduce the results. To facilitate this, each chapter is split roughly into two parts, "The

Atomic Pair Distribution Function Analysis. Simon J. L. Billinge and Kirsten M. Ø. Jensen, Oxford University Press.

Problem" and "The Solution". The first lays out the scientific question and details of the experiments carried out and a brief discussion of the approach taken to analyse the data. The reader is encouraged to download the data and structural models from the internet for this problem and attempt their own analysis. There is a section called "What next?", which lays out some steps to try on your own, and a section "Wait, what? How do I do that?", which lays out in more detail how to do those steps if you need some hints, but without giving away how to get at the scientific answers. There will be a "Problems" section that asks some questions designed to get you thinking about the link between the data and the scientific questions, and highlighting some "gotchas". In the "Solution" section of the chapter, we present our solution to the problem, which may be used by the reader as a crib if you get stuck or to check your solution against. Hopefully you manage to do as well as we did, or even better!

Our goal is to help new users get to grips with PDF analysis and build confidence in their abilities by finding their own way to the right solution. In this way you can build up valuable intuition and expertise that you may then apply to your own scientific problems. The book attempts to provide an answer to the question "I am excited to try PDF, how do I get started?". With this in mind we have concentrated on methods based on model-free analysis of PDF data and relatively simple modelling approaches such as real-space Rietveld using the PDFGUI program, described further below. Since the complex modelling program, DIFFPY-CMI (CMI stands for complex modelling infrastructure) is growing in power and utility, we also include some examples using this program. This program has a somewhat steep learning curve unless you are familiar with using the Python programming language, and we do not recommend it initially. However, a good way to get over the CMI learning curve is to repeat (and extend) some of the simple examples that you have already done using PDFGUI. To that end, we provide CMI solutions to all the PDFGUI examples we give and some more problems that cannot currently be done with PDFGUI. We do not venture into more advanced and computationally expensive modelling such as big box models based on reverse Monte Carlo methods (Mcgreevy *et al.*, 1990; McGreevy, 2001; McGreevy and Zetterstrom, 2003), potential based modelling (Soper, 1996; Hwang *et al.*, 2012), nanostructure solution (Juhás *et al.*, 2006; Cliffe *et al.*, 2010), or emerging complex modelling (Billinge, 2010; Tucker *et al.*, 2007; Krayzman *et al.*, 2009; Farrow *et al.*, 2014) approaches that combine multiple heterogeneous datasets in a global optimization scheme. The reason is not because these approaches are not useful – they are. However, our main goal is for the reader to build intuition and confidence in interpreting features in the PDF pattern, and this is best done with simple models, or no models at all, and often the scientific question may be answered without the more complicated approaches!

Of course, our approach to the book also naturally lends itself to us using our own scientific studies as the problem set, since we have the data (and we know the solution!). There are many many exciting and scientifically insightful studies by other authors in the literature, and we do not mean to imply that the ones chosen are in any way superior – just more conveniently to hand.

This approach also allows us to point out and discuss some of the gotchas and trip-wires that can be so frustrating for new users and to have a few rants about what we think are common misconceptions in the community.

1.3 Why PDF?

We are assuming you are reading this because you have already heard about PDF and what it is for, and want to get started. However, to find out a bit more about this, please see Chapter 2. Briefly, PDF is a method for finding the local and intermediate-range atomic structure in materials. Our lives have been revolutionized by the development of crystallography, which is a method for finding the arrangement of atoms in a crystal. But what if the material that you are interested in is not a crystal but is a nanoparticle, or an amorphous material, or a mixture of those, or some other complex multi-structured material? In these cases you may want to try using the PDF method to find where the atoms are and how they are arranged, and if you are interested in trying out PDF for this, then you have come to the right place! As we mention above, if you want to find out more about how and where PDF has been applied to solve real-material problems, beyond the chapters in this book, some good overview articles include Keen and Goodwin (2015), Young and Goodwin (2011), Billinge (2008b), Billinge and Kanatzidis (2004), and Lindahl Christiansen *et al.* (2020a).

1.4 Software

To do the problems in the later chapters you will need some software. The choice of software to use is entirely your own. However, the worked solutions will be limited to specific programs. We list below the software that we use in the chapters and give some guidance about where to get it. It is important to understand that this book is not intended as a software manual, and we do not reproduce here the contents of the manuals. Please refer to the web pages and the manuals that come with the software to get it installed and working. It is advised that you then follow some of the step-by-step tutorial examples that come with the software before moving on to the examples in this book. This book is designed to pick up where those software-manual tutorial examples leave off.

Chronologically, after doing total scattering measurements, the first thing that you do in a PDF project is to reduce raw data to a PDF. Depending on your measurement, this can include data integration and corrections followed by a Fourier transformation. However, the focus of the book is to obtain scientific information from the PDF, and so we will begin by describing the PDF *modelling* software. Most of the examples in this book take as a starting point a PDF obtained from properly reduced data from a scattering experiment, and so it is possible to follow (nearly) the whole book without doing battle with the data reduction part. As time goes on the data reduction is becoming more automated, and you may well already have returned from a beamtime with PDFs provided by an instrument scientist if it was a synchrotron x-ray or a neutron experiment. However, despite leading off with the modelling software, to be a full, card carrying PDF expert you will have to become expert in the data reduction part as well, so in Chapters 4, 10, and 12 we also give diffraction data processing examples for you to play with.

In the following sections, we describe the PDF-specific programs recommended for the problems in the book. Apart from these programs, it is also a good idea to have software for data visualization ready, i.e. a plotting program of your own choice. This could be Origin, MagicPlot, or even Microsoft Excel. If you like using scripts, both Matlab or Python have excellent plotting tools. PDFGETX3, which we describe below, includes a plotting package PLOTDATA for plotting in Python.

We also recommend having a structure visualization program installed so that you can inspect the structure files (mostly .cif files) that we provide with the data. This could be VESTA (Momma and Izumi, 2011), CRYSTALMAKER (Palmer, 2015), MERCURY (Macrae *et al.*, 2020), or similar.

1.4.1 Quick start

The only PDF software you need to get started with the examples in this book is PDFGUI. This is a GUI (graphical user interface)-driven program for PDF modelling. Go ahead and download it from diffpy.org, where you will also find instructions for installation. If you just want to get started right away, that is it for the software introduction. You can ignore the rest of this chapter. However, if you want to do more fancy things, or if you run into trouble with your installation, we give some background below (and share some of our prejudices) that can help you get further. Finally, it is also *strongly* advised to join the Diffpy-users Google group (https://groups.google.com/forum/#!forum/diffpy-users) where users post questions and the community post answers about the DIFFPY software (used here), and PDF analysis in general. Many of your questions may already be answered in the Diffpy-users group, but if not, post your question, get it answered, and the answer will be shared with the whole community.

1.4.2 Basic computational infrastructure

All the PDF software we describe and use in this book is Python based. If you are using PDFGUI only, you do not need to know very much about Python to get started. However, Python and the various tools that make the Python ecosystem work are the tools of choice for scientific programming right now. Time spent getting a bit of familiarity with Python (and its ecosystem) is time well spent. If you are interested in knowing more, we have put some information about getting started in the Pythonic world into Appendix A, which can also help you if you run into trouble installing Python-based software such as PDFGUI and DIFFPY-CMI and running Python scripts.

1.4.3 Data modelling and interpretation software

As we will see throughout this book, the PDF contains a wealth of information that can be deduced without any modelling. This can be just making intuitive deductions from positions and widths of peaks, to model independent quantitative analysis of the data such as described below using the PDFITC website. But at some point it is always beneficial to build atomistic models, compute their PDFs, and compare these to the measured PDFs. When we do this there are parameters in the models, and we can then vary these parameters in a regression loop (fitting) in such a way as to give the best fit of the PDF calculated from our model to the measured PDF. This process is also widespread in fitting single crystal and diffraction data from powdered crystals (in the latter case it is called "Rietveld refinement"). In principle, one could come up with any kind of representation for the structure and figure out a way to parameterize it and compute the PDF. The most common is to put "atoms" into a "box" and change the size (and shape) of the box and move the atoms around. If the structure of the thing being modelled is reasonably well ordered, and is uniform, a crystallographic representation of the structure is often sufficient, and the "box" is the unit cell. In more complicated cases we may need to make the box larger and make "supercells" of the unit cell to capture broken symmetries in the local structure that are not there in the average structure. We can take this box-enlarging game to any level and make large boxes with many thousands of atoms in them, which is the basis of "big-box" modelling such as reverse Monte Carlo methods, which are not addressed in this book.

1.4.3.1 PDFITC

PDFITC stands for "PDF in the cloud". It is a website at the URL `pdfitc.org` that contains a set of software applications that run on cloud computing and are for interrogation of PDF (and some powder diffraction) data. As time goes on, more apps will be added, but at the time of writing the website contains STRUCTUREMINING, SPACEGROUPMINING, PEARSONMAPPING, and NMFMAPPING. The workflow for the early apps is all the same. Either one, or multiple, experimental (or simulated, actually) PDFs are uploaded to the app with a minimal amount of metadata. The app does some calculations, and a list of results files are returned to the user. The apps are largely intended for early interrogation of the data. For

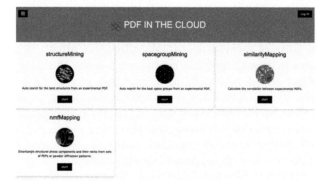

Figure 1.1 Screenshot of the PDFITC PDF analysis website.

example, STRUCTUREMINING will return a list of structures (and their CIFs) that it thinks might be good candidates for more detailed fitting with PDFGUI. SPACE-GROUPMINING uses a supervised machine learning model and, given a PDF (and no other metadata), it returns what it thinks is the spacegroup of the material that was measured. Both of these are doing statistical modelling, and just as with a Google search, the top returned answer may not be the best, but the hope is that the right answer is near the top, if not at the top, and that the other top ranked answers are also interesting and closely related in some way. The hope is that this can save the user a great deal of time early in the analysis of a dataset and also lead to better science by finding lines of inquiry that may not have been expected.

The PEARSONMAPPING and NMFMAPPING apps have a slightly different objective. They take as input sets of data (possibly hundreds or thousands of PDFs/patterns), and they use statistical methods to find hopefully interesting information. PEARSONMAPPING computes the pairwise Pearson correlation coefficient parameters between each dataset and returns the matrix of the results. It allows you to find similar datasets and to see exactly where in a series of data some aspect of the signal is changing. It is a visual way of rapidly looking for phase transitions, for example. NMFMAPPING uses non-negative matrix factorization, an unsupervised machine learning method, to try and explain the set of data in a small number of components. For example, a set of data from an *in situ* synthesis experiment might have chemical components coming and going with time. NMFMAPPING will try and extract the PDF signals of each of the chemical components and how much contribution each component is making towards each measured signal. In PDFITC, the NMFMAPPING app returns the component PDFs and the weights.

In the workflow of PDF analysis activities, the use of PDFITC lies after the data reduction from raw data to the PDF and before the more detailed PDFGUI and DIFFPY-CMI modelling activities.

1.4.3.2 PDFGUI

PDFGUI is the main PDF modelling software program that we will use in this book. It grew out of its PDFFIT predecessors (Proffen and Billinge, 1999; Billinge, 1998), and with some investment by the US National Science Foundation (NSF) through the DANSE software development program, a graphical user interface (GUI) was developed (and the engine rewritten), as shown in Figure 1.2.

The program mimics a crystallographic refinement approach such as Rietveld refinement. It uses a crystallographic representation for the structure in terms of atoms at fractional coordinates within a unit cell that is in a crystallographic coordinate system. This makes it very easy to build models if you have a CIF (crystallographic information file) from a structural database, for example. When you do PDF analysis, you are generally interested in structural features that go beyond the average, crystallographic structure, in other words, some local symmetry breaking or distortions. This is easily accomplished by, for example:

- Using a different (generally lower symmetry) space-group than that of the crystal structure

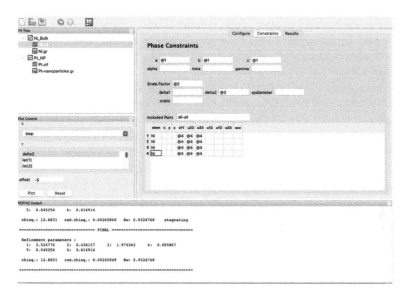

Figure 1.2 Screenshot of the PDFGUI structure fitting program.

- Starting with the crystallographic space-group but allowing atoms to move off special positions, or removing constraints coming from the symmetry
- Making supercells that allow ordered patterns of displacements to be described

There is no contradiction between having a crystallographic representation for the structure and refining structures to non-crystallographic objects.

The crystallographic nature of the PDFGUI design does have limitations. For example, because it uses periodic boundary conditions, it is impractical to build models for non-uniform structures such as small nanoscale clusters of atoms or highly disordered layered materials. However, there are many "tricks of the trade" that allow you to build approximate models for situations such as this that we will discuss in later chapters. This means that getting insights into scientific questions about even quite complex situations is possible using this simple small-box modelling. PDFGUI is freely available as it was funded by NSF, and it can be downloaded from diffpy.org. If you have issues, check Appendix A for some Python instructions.

1.4.3.3 DIFFPY-CMI

It is possible to carry out much more complex fits than possible in PDFGUI by using DIFFPY-CMI. At the time of writing, there is no GUI for DIFFPY-CMI. Even when it is developed, the new GUI will remove some of the powerful flexibility of DIFFPY-CMI. To get the most out of DIFFPY-CMI it is necessary and encouraged to learn a few Python programming skills. In this book we provide DIFFPY-CMI solutions to all the problems, whether they can be solved in PDFGUI or not, so you can practice your DIFFPY-CMI skills and adapt the scripts to your own scientific problems. For each chapter there is a DIFFPY-CMI script in the solutions folder, along with the PDFGUI solution. In some of the later chapters, the problems are more advanced and cannot be solved using PDFGUI at all, in which case the solution only involves DIFFPY-CMI scripts.

A new version of PDFGUI, PDFGUI3, is under development which reuses the PDFGUI GUI that we know and love, but uses a DIFFPY-CMI engine under the GUI. In later releases the PDFGUI3 GUI will be extended to expose in the GUI different functionalities that are currently now only available in DIFFPY-CMI. However, there will always be functionality in DIFFPY-CMI that is not present in the GUI, so learning how to script using DIFFPY-CMI in Python will still be a useful skill. DIFFPY-CMI is almost infinitely extendable (it was designed with this in mind), and so we hope that the more programming-savvy among you will contribute code to add your extensions to future DIFFPY-CMI releases and make them available to the whole community. This can be done by making Pull Requests into the diffpy GitHub project (https://github.com/diffpy). If you are not so software-experienced but want a new feature or capability, you can request it by posting an issue at the same place or making the request on diffpy-users.

For instructions on downloading and installing DIFFPY-CMI, please go to diffpy.org, and remember to check out Appendix A for some Python tips and tricks.

1.4.4   Data reduction software

As we will describe in Chapters 2 and 14, there are many different ways of obtaining scattering data for PDF analysis. Depending on the measurements, there will be different steps in the data reduction process, including instrument calibration, data correction, data integration, and the final Fourier transformation. The most commonly used technique for PDF measurements, also used for obtaining data for most of the chapters in this book, is the rapid-acquisition x-ray PDF mode (Chupas *et al.*, 2003), which can be used at high-energy synchrotron beamlines. We describe this technique in Chapters 2 and 14, and how to get a PDF from it in Chapter 4. Here, we will just briefly introduce software used for obtaining the PDF from integrated x-ray scattering data.

1.4.4.1   *PDFGETX3 and XPDFSUITE*

PDFs are obtained from properly integrated, normalized scattering data corrected for all other effects than coherent, elastic scattering. We describe what that means in a bit more detail in Chapter 2. Assuming you have a properly integrated 1D powder diffraction pattern from your sample, you therefore need to carry out corrections for experimental effects, such as removing parasitic scattering such as backgrounds and fluorescence, and correcting for multiplicative effects such as sample absorption. You will then need to scale the data and divide by the atomic form factor. Finally, if you want to see the real-space PDF function you will need to Fourier transform the data to convert it from reciprocal-space to real-space. In this book, we use our most recent program PDFGETX3 (Juhás *et al.*, 2013), which is fast and easy to use. Getting all the corrections precisely correct is an arduous and detailed process. Readers are referred to Chapter 5 of (Egami and Billinge, 2012) for all the gory details. However, an *ad hoc* approach to the data reduction was recently proposed (Billinge and Farrow, 2013) and implemented in PDFGETX3. This gives comparable results to the full analysis (Juhás *et al.*, 2013), for example, carried out in PDFGETX2 (Qiu *et al.*, 2004), and is much much easier to use. PDFGETX3 is free for academic research. Instructions for obtaining PDFGETX3 may found at diffpy.org.

PDFGETX3 has excellent functionality but does not have a full-featured graphical user interface (GUI). To address this, and to help with the entire end-to-end workflow of processing and then modelling PDF data, we developed XPDFSUITE. This greatly increases productivity, especially when dealing with large numbers of datasets, and we strongly recommend that you consider using it. It is a commercial product with licences for academic research and for commercial applications, and instructions for obtaining it are available through diffpy.org.

Other programs, from our and other groups, for these steps include PDFgetX2 (Qiu *et al.*, 2004), and GUDRUNX (Soper, 2012). Equivalent programs exist for processing time-of-flight neutron data, such as PDFgetN (Peterson *et al.*, 2000) and GUDRUN (Soper, 2012). There is also now a version of PDFGETN3 for neutrons that uses the same algorithm as PDFGETX3 but applied to neutron data (Juhás *et al.*, 2018). Most time-of-flight neutron sources that are suitable for PDF work will send you home with data transformed to $S(Q)$ and $G(r)$ ready for modelling, so we do not give examples of how to use these tools here. Chapters that give examples of modelling neutron-PDF (nPDF) data will assume that you already have PDFs.

2

PDF primer

Kirsten M. Ø. Jensen and Simon J. L. Billinge

2.1 Introduction

Coming from a background such as chemistry, materials, or earth science, chances are that you have already been introduced to basic crystallography and x-ray scattering. This gives you a good background for starting PDF analysis, as the PDF world builds on a lot of the same ideas used in other scattering techniques. Throughout the book, we therefore assume that you are familiar with fundamental concepts such as the Bragg law, unit cells, lattice symmetry, and space groups, which are covered in most undergraduate solid state chemistry or materials science courses. However, before diving into PDF data treatment in the following chapters, we will give a short introduction to the PDF and how it is obtained from x-ray scattering data. This chapter is only meant to get you started and will not cover scattering or PDF theory – for that, we refer you to *Underneath the Bragg Peaks* (Egami and Billinge, 2012) or other scattering textbooks. Instead, we will simply connect the dots between atomic structures, scattering signals, and the PDF, at the level you will need when going through the chapters in the book. In this chapter we will focus on x-rays and only introduce neutron and electron PDFs briefly, although these are both important probes for scattering.

2.2 X-ray scattering from materials

X-ray diffraction is all about interference effects. You have probably already learned this when deriving the Bragg law, where the crystal lattice is described in terms of lattice planes that can reflect x-rays. The Bragg law shows that a certain relation between the x-ray wavelength λ, the diffraction angle 2θ, and the separation between crystal lattice planes d must be fulfilled in order to see Bragg peaks: $2d\sin(\theta) = n\lambda$. However, the point of many PDF experiments is to get insight into structures *without* long range order (and thus lattice planes), for example, nanoclusters, particles, or local deviations away from an average crystal structure. In order to understand how this information can be obtained from scattering data,

Atomic Pair Distribution Function Analysis. Kirsten M. Ø. Jensen and Simon J. L. Billinge, Oxford University Press.
© Simon J. L. Billinge and Kirsten M. Ø. Jensen (2023). DOI: 10.1093/oso/9780198885801.003.0002

it is important to remember that the Bragg law, while incredibly useful, does not quite describe the actual physical scattering processes taking place in the sample.

X-rays interact with each of the electrons in the materials in the sample, no matter if it is a crystal or an amorphous compound. This interaction happens through several different mechanisms, but for us, the most important one is what is known as *Thomson scattering*, in which the incoming x-ray beam is scattered elastically from the electron, so that the x-rays leaving the electron have the same wavelength as the incoming beam. As the waves scattered from each electron meet each other, they interfere, and a scattering pattern arises. The pattern is characteristic of the arrangement of electrons in the sample, and through data analysis, we can extract this information and construct an atomic structural model. In crystalline materials, the atoms, and thus the electron clouds, are arranged in particularly simple periodic patterns, which gives rise to well-defined constructive interference effects in the scattering pattern, i.e. the Bragg peaks that we know and love. However, any other atomic arrangement will also give rise to interference effects, which we can refer to as *diffuse scattering*. By analysing the diffuse scattering, we can get information on the disorder in a structure.

2.2.1 Collecting scattering data from a sample

In an x-ray scattering experiment, we collect data by measuring the intensity of the scattered x-ray beam at different angles from the incoming beam, as illustrated in Figure 2.1 showing a measurement geometry very often used for PDF data collection, also known as the "*RA-PDF*", Rapid Acquisition PDF, setup (Chupas *et al.*, 2003).

Here, the use of a large 2D detector means that we can cover a wide angular range of both 2θ and ϕ in a single measurement. Figure 2.2 shows examples of such 2D x-ray scattering patterns from a powder of a highly crystalline material (CeO_2) and a nanostructured material (1.5 nm gold nanoparticles), where we see how the nanoparticles result in broad diffuse features compared to the sharp Bragg peaks from the crystalline sample.

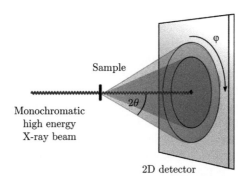

Figure 2.1 *RA-PDF* setup used for x-ray total scattering data collection.

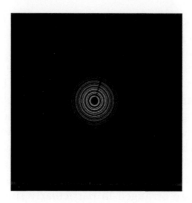

Figure 2.2 2D scattering patterns collected for bulk CeO_2 (left) and gold nanoparticles (right).

The structural information in both patterns is *isotropic* as the scattering intensity appears identical for all values of the azimuthal angle ϕ. On closer inspection, this is not completely true because of experimental effects such as beam polarization, but these non-idealities are corrected by the data analysis software. What is important is that there is no structurally relevant information in the ϕ dependence from an isotropic sample. In Figure 2.2(a) and (b), the samples are powders, i.e. they consist of many small particles present in an ensemble in which all particle orientations are equally likely. For such isotropic samples, which also include solutions, liquids, and glassy and amorphous materials, the nature of the scattering pattern means that we only have to consider the scattering pattern as a function of the scattering angle 2θ. We can therefore integrate the scattering pattern around the azimuthal angle, ϕ and obtain 1D diffraction patterns as illustrated in Figure 2.3. This is in contrast to single crystal diffraction, or scattering from highly textured samples, where we have to take the azimuthal angle, ϕ, into account and deal with data in 2D. The PDF of samples with texture has been studied (Gong and Billinge, 2018; Cervellino and Frison, 2020b; Harouna-Mayer *et al.*, 2022) but is not discussed in this book. Throughout the book, all the samples we will work with are isotropic.

You may have noticed that the integrated scattering patterns are plotted as a function of Q in Å^{-1}. In standard powder diffraction, 1D scattering patterns are often plotted as a function of the scattering angle 2θ. However, this will give different patterns from the same material if they were measured with different-wavelength x-rays or neutrons. It is therefore preferred to use a quantity that depends only on the structure and not on the measurement. Sometimes the magnitude of the crystallographic d-spacing is used, but in the world of PDF, it is generally preferred to use Q, the magnitude of the scattering vector, $|\vec{Q}|$, instead, From a physics perspective, $\vec{Q}$ is also the momentum transfer in the scattering process. It is illustrated in the vector diagram in Figure 2.4. Thomson scattering is elastic, i.e. there is no change

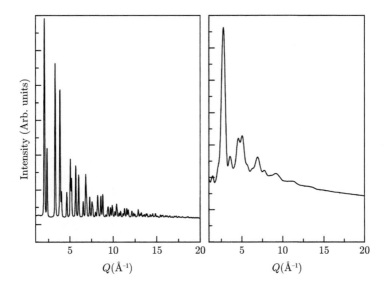

Figure 2.3 Integrated 1D scattering patterns collected for bulk CeO$_2$ (left) and gold nanoparticles (right). These 1D patterns were obtained by integrating azimuthally around the concentric rings of scattering shown in the 2D intensity images shown in Figure 2.2.

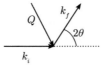

Figure 2.4 Vector diagram defining $\vec{Q}$.

of beam energy in the scattering process. This means that the length of the two $\vec{k}$-vectors are equal, and we can derive the relation between the magnitude of the vector, $Q = |\vec{Q}|$ and $\theta = 2\theta/2$ from simple geometry,

$$Q = 2k\sin(\theta) = 4\pi\sin(\theta)/\lambda. \tag{2.1}$$

While most software for treatment of powder diffraction takes scattering data as a function of 2θ, the PDF world (and physicists in general) mostly consider data as a function of Q. We will do the same throughout the book.

2.2.2 Relating scattering patterns to atomic structure

So how do scattering patterns appear, and how do they depend on the atomic arrangements in the sample? As described above, the scattering signal arises as the waves scattered from each electron in the sample interact with each other. While we will not do it here, it takes just a bit of maths to show that for a structure that contains N atoms at positions $\vec{r_i}$, the $\vec{Q}$-dependent amplitude of the scattered wave becomes:

$$\psi\left(\vec{Q}\right) = \sum_{i=1}^{N} f_i(\mathrm{Q}) \exp\left(i\left(\vec{Q} \cdot \vec{r_i}\right)\right). \tag{2.2}$$

Here, $f_i(Q)$ is the atomic form factor for x-ray scattering, which is dependent on the number of electrons in each atom and the spread of the electron cloud around the atom.

Equation 2.2 is completely general, and it holds for crystalline, nanocrystalline, and amorphous materials for the calculation of scattering amplitudes. Quite often, it is not very practical though, as the sum is over *all* atoms in the sample. For crystalline materials, where the scattering amplitude is non-zero only at $\vec{Q}$–values corresponding to Bragg peak positions, it can be much simplified so that we only need to sum over the atoms in the crystallographic unit cell, giving us the well-known crystallographic structure factors.

Given a known structure consisting of N atoms with positions $\vec{r_i}$, we can calculate the scattering amplitude. Does that mean that we can directly take the measured data and get the atomic positions? Unfortunately not. When collecting data, we cannot measure the scattering amplitude but only the scattering intensity (proportional to the squared amplitude), which means that we lose valuable information about the absolute phase of the interfering waves. In standard crystallography this is known as the phase problem. However, for crystalline materials, and especially in single crystal diffraction, many methods have been developed to overcome this issue, and in many cases, structure solution can be done routinely. For nanomaterials, structure solution is still not a solved problem (Billinge and Levin, 2007). However, PDF analysis of scattering data is a good tool to get closer to high quality quantitative structural models, as we will see throughout this book.

If the sample, as in our case, is isotropic, we can simplify Eq. 2.2 and consider scattering only as a function of Q, i.e. vector length. Performing this simplification, we get the Debye scattering equation for scattering intensity, derived in 1915 by Pieter Debye (Debye, 1915),

$$I\left(Q\right) = \sum_i \sum_j f_i\left(Q\right) f_j\left(Q\right) \frac{\sin(Qr_{ij})}{Qr_{ij}}. \tag{2.3}$$

In this equation the sum is taken over all pairs of atoms i and j, and r_{ij} is the distance between atoms i and j.

2.3   Obtaining the PDF from x-ray total scattering data

We obtain the PDF by Fourier transforming the measured scattering intensity. While we will not go much into it here, the Fourier transform is an incredibly useful mathematical tool that separates a signal into individual frequency contributions. In our case, it takes the measured scattering intensity and separates this into peaks relating to the positions of the electrons (and thus atoms) that the x-ray scattered from in the first place: the PDF.

However, before we Fourier transform the data, corrections to the measured intensities are needed. As mentioned briefly above, we have so far only considered Thomson scattering, which is an elastic, coherent scattering process. Other events will also take place as the x-ray beam interacts with the sample, including Compton scattering (which is another scattering process, but inelastic and incoherent), absorption of the beam by the sample, fluorescence, multiple-scattering (i.e. the beam is scattered more than once before exiting the sample, scrambling the structural information), and so on. Furthermore, the measured data may contain a signal from the sample holder and from air in the x-ray beam path. Before the Fourier transform can be done, the data must be corrected for all these effects so that we isolate the elastic, coherently scattered signal from our sample. This can be done either explicitly by calculating all the contributions and subtracting them from the data (see the mammoth Chapter 5 of (Egami and Billinge, 2012)) or by using the much simpler *ad hoc* approach, as we will discuss further in Chapter 4.

After all corrections are done, the scattering amplitude is normalized to obtain $S(Q)$, the total scattering structure function, which is shown for the CeO_2 and gold nanoparticle samples in Figure 2.5. $S(Q)$ is obtained by dividing by the number of scatterers N and the average scattering power per atom, which for x-rays is given by the square of the atomic form factor, $f(Q)$,

$$S(Q) = \frac{1}{N \langle f \rangle^2} (I(Q) + \langle f \rangle^2 - \langle f^2 \rangle). \tag{2.4}$$

Here we have left out the Q-dependence of f for simplicity. The additional terms, $\langle f \rangle^2 - \langle f^2 \rangle$, represent the Laue monotonic diffuse scattering that is subtracted to obtain the correct normalization, such that the average value $\langle S(Q) \rangle$ is 1.

From $S(Q)$, we can easily calculate the reduced total scattering structure function,

$$F(Q) = Q(S(Q) - 1), \tag{2.5}$$

which is often referred to in the PDF literature and is plotted for our two samples in Figure 2.5.

By now taking the sine Fourier transform of $S(Q)$, we obtain the reduced pair distribution function,

$$G(r) = \frac{2}{\pi} \int_{Q_{min}}^{Q_{max}} Q(S(Q) - 1) \sin(Qr) \, dQ. \tag{2.6}$$

In principle, the integration over Q in 2.6 should be done from zero to infinity. This is of course not physically feasible, as the experimental configuration dictates both Q_{min} and Q_{max}. Eq. 2.1 shows that these are dependent on the 2θ range covered by the detector and by the x-ray wavelength. Terminating the integration at a finite Q-value results in a broadening of the PDF peaks and oscillations in the Fourier transform, known as termination ripples. These are evident in Figure 2.6, where PDFs from the CeO_2 data are calculated using two different Q_{max} values,

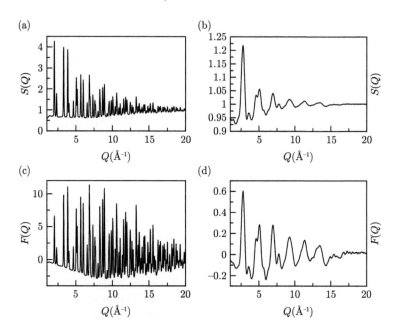

Figure 2.5 $S(Q)$ for bulk CeO$_2$ (a) and gold nanoparticles (b). $F(Q)$ for CeO$_2$ (c) and gold nanoparticles (d).

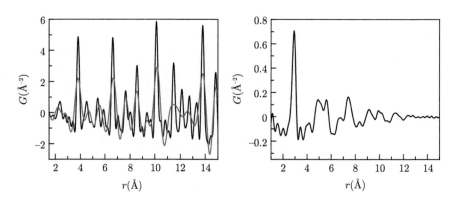

Figure 2.6 Left: $G(r)$ for bulk CeeO$_2$ calculated with Q_{max} 20 Å^{-1} (black) and 10 Å^{-1} (red). Right: $G(r)$ for gold nanoparticles, $Q_{max} = 20$ Å^{-1}

10 and 20 Å^{-1}. By using a high value for Q_{max}, we obtain a higher resolution in r-space, i.e., interatomic distances: in the material that are more similar can be better resolved, and smaller termination ripples, allowing us to extract more information from the data. X-ray scattering experiments for PDF analysis are therefore usually done with a high x-ray energy (often between 50 and 90 keV) and with the

detector placed close the sample, so that a wide angular range is collected on the detector, giving Q_{max} values of 20 to 30 Å^{-1}. We refer to this as a "total scattering" experiment both because of the wide Q-range and because we consider both Bragg and diffuse scattering in the following data analysis. In a standard powder diffraction experiment for Rietveld refinement, the data are often collected to much lower Q_{max} values somewhere between 6 and 12 Å^{-1}.

We should note here that extending the Q-range increases the chance of including noise in the data, as the x-ray scattering power of atoms decreases with Q. Determining the optimum Q_{max} to use in the Fourier transform is therefore often a compromise between r-resolution and noise level. We will look more closely into this in Chapters 4 and Chapter 10.

Using the set-up illustrated in Figure 2.1, placing the detector close to the sample, has one disadvantage, as the total scattering pattern is "squeezed" into the limited space of the detector, which has a finite number of pixels. This can result in increased instrumental broadening due to geometric and pixelation effects. For crystalline materials, Bragg peaks will be broader and will overlap. When collecting conventional powder x-ray diffraction (PXRD) data for Rietveld refinement using a similar detector, a much smaller 2θ-range is collected, which gives less peak overlap, meaning that Bragg peak positions (and thus unit cell parameters) may be better determined from a Rietveld experiment than a PDF experiment. It is important to keep this in mind and to consider which technique and data collection strategy best suits your scientific problem.

2.4 The pair distribution function

The Fourier Transform in Eq. 2.6 gave us the *reduced pair distribution function*, $G(r)$ from x-ray total scattering data. $G(r)$ is a member of a large family of PDF functions, as many different related functions with only subtle differences in their definitions and units are used by different programs and communities. Basically, they all contain the same information, a histogram of interatomic distances in the sample. Before further discussing $G(r)$, we first introduce $R(r)$, the radial distribution function, which is physically very easy to interpret. $R(r)$ is given by,

$$R(r) = \frac{1}{N} \sum_{v} \sum_{u} \frac{f_v f_u}{\langle f^2 \rangle^2} \delta(r - r_{vu}), \tag{2.7}$$

where r_{vu} is the distance between the v-th and u-th atom in the sample, and f_u and f_v are the Q-dependent atomic form factors of the elements. We sum over all the atoms in the structure. The PDF peak positions directly give the distance between atoms in the sample, and their intensity is dependent on the number of electrons in the atomic pair (through the f_u and f_v dependence), as well as the number of atomic pairs of the given distance. $R(r)$ is illustrated for a simple crystalline single atomic structure in Figure 2.7.

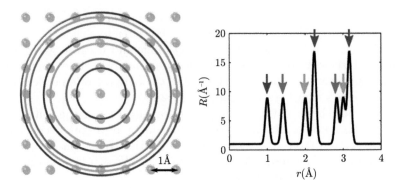

Figure 2.7 $R(r)$ for a simple single atomic 2D crystalline structure. The coloured circles show interatomic distances from the central atom, corresponding to the peaks in the PDF. The PDF peak intensity is for the single atomic structure given only from the number of atomic pairs with a certain distance between them in the structure.

The relation between $G(r)$ and $R(r)$ is given by:

$$G(r) = \frac{R(r)}{r} - 4\pi\rho_0 r, \tag{2.8}$$

where ρ_0 is the number density, i.e. the number of atoms in a volume of the sample.

While $R(r)$ is more intuitive than $G(r)$, it is much more common for us to use $G(r)$ when doing PDF analysis. This has several reasons. Most importantly, $G(r)$ is the function we obtain directly from the Fourier transform of the processed experimental data. If we are to calculate $R(r)$ (or any of the other related PDF functions), we need to make assumptions about the sample structure, such as the atomic number density and the particle size and shape, which we do not necessarily know. By using $G(r)$ directly, all assumptions on the sample lie in the model itself, which ensures a cleaner separation between data treatment and structural modelling. Furthermore, as seen in Figure 2.8, which shows $R(r)$ and $G(r)$ for a Ni sample, $R(r)$ diverges as r^2 with increasing r. If we think about the number of neighbours at some fixed distance around an atom at the origin, we realise that as r increases, the number of atoms will increase and increase, because the size of a spherical annulus at that r becomes larger and larger. From a practical point of view, this makes the $R(r)$ function very inconvenient to visualize. In contrast, $G(r)$ oscillates around 0, and the amplitudes of the PDF peaks give a direct measure of the structural coherence in the sample. If a sample had a completely random distribution, $G(r)$ would be a flat line. If a certain interatomic distance is present frequently (i.e. more than average), a positive peak is seen in $G(r)$, while an interatomic distance present less than average will give negative values for $G(r)$. Finally,

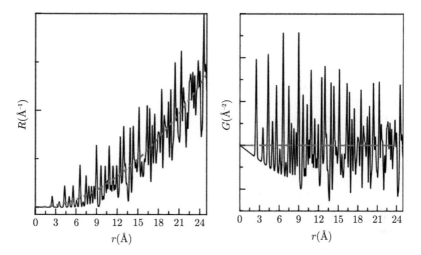

Figure 2.8 Left: The $R(r)$ function for Ni (black) and $4\pi r^2 \rho_0$ (red), showing the average density for a material with constant number density. Right: The corresponding $G(r)$ function, oscillating around 0.

in the $G(r)$ function the experimental uncertainties are approximately constant in r, which also leads to more reliable modelling and visual interpretation of the data.

2.5 Extracting structural information from the PDF

The PDF contains all information about both long- and short-range order in the sample, as does $S(Q)$. However, since it is now a function in real space (r-space, instead of reciprocal-space), the data analysis is very intuitive: the PDF simply represents a histogram of all interatomic distances in a sample. The peak positions tell us about the interatomic distances between pairs of atoms in the sample, which for the first PDF peaks (at lowest r values) can be directly related to bond lengths in the material. The peak intensity is related to the number of pairs of a certain type as well as the atomic number of the elements in the pair, and it can be used to deduce coordination numbers in the structure, or chemical ordering. The peak width tells us about the positional disorder of atoms in the structure and relates to the thermal motion of atoms. The range of the structural coherence in the sample is also contained in the PDF as the value of r when the signal dies out. Due to these intuitive qualities, much information can be extracted from a simple, model-free approach to PDF data analysis where information about bond distances and coordination numbers may be inferred visually or can be found by simple Gaussian peak fitting. Furthermore, a great deal more may be learned about the structure of a sample from comparison of experimentally determined PDFs with control samples or PDFs calculated from a known structure. Model-free analysis of time- or temperature-dependent PDFs can furthermore be used to follow structural changes by tracking the change in peak position and width. We will discuss this kind of model-free analysis throughout the book, and we describe specific tools and software packages for it in Chapter 15.

While model-free PDF analysis can take us far, the most information can be extracted from the PDF through structure modelling. By calculating the PDF from a model structure and "refining" its parameters until a good agreement between the calculated and experimental PDFs is obtained, a structure describing the sample can be found. As already mentioned in Chapter 1, PDF structure modelling can be done in many different ways, varying in complexity and time consumption, and using different approaches such as "big box modelling", or "small box modelling", which we focus on in this book. The refinement starts from a known structure, which is expected to be quite close to the final model.

One small-box method is the so-called "real-space Rietveld refinement", which can be used in PDFGUI or DIFFPY-CMI. This approach can be used for materials where the structure is expected to be closely related to a known crystal structure and thus contain some long-range or medium-range order. By building up a model structure based on a crystallographic unit cell and refining the parameters in the model (lattice parameters, atomic fractional coordinates, atomic displacement parameters, etc.) using a least-squares fitting routine, a structure for the sample can be extracted. This is quite similar to doing a classical Rietveld refinement of PXRD data, but unlike a conventional Rietveld refinement (where only the intensity present in Bragg peaks is taken into account), the model can catch local structural phenomena, such as local disorder from the average crystal structure, as already mentioned in Chapter 1. For small nanoparticles, where Bragg peaks are either very broad or almost non-existent, real-space refinements can often give a much more accurate structural model than a classical Rietveld refinement in reciprocal-space. This is done by multiplying the PDF calculated from a crystal structure with an envelope describing the extent of structural coherence. In a similar manner, real-space Rietveld can even help in deducing structural motifs in amorphous materials.

Another approach to small box modelling is building finite clusters of atoms and computing the PDF. This can be done by computing the scattering pattern using the Debye scattering equation and Fourier transforming it. Unlike the real-space Rietveld method, a structure refinement using the Debye equation is not based on a unit cell and does not assume any structural periodicity in the sample. Instead, the model consists of a structure with a discrete number of atoms, all with well defined coordinates. We introduced the Debye scattering equation in Eq. 2.3, which shows how the scattering intensity, and thus the PDF, can easily be calculated from a discrete structure. By refining the parameters in the model, e.g. the atomic displacement parameters of the atoms, a final model for the sample can be obtained. This approach is very useful for refining the structure of nanoclusters and molecular complexes, as long as a good starting model is used and care is taken when determining the refinement parameters. PDFGUI does not support structural refinement of discrete structural models without translational symmetry, but such analysis can be done in DIFFPY-CMI.

2.6 Measurement of total scattering data

2.6.1 PDFs from x-rays

Above, we have already mentioned a few aspects of how we measure x-ray scattering data for PDF analysis. In Figure 2.1, we showed the widely used RA-PDF setup: using high-energy x-rays (usually 60–90 keV) from a synchrotron and a large 2D detector placed close to the sample, we are able to rapidly collect scattering data to large Q values with statistics sufficient for PDF analysis, and data can be collected on a second, or even millisecond, time-scale. This kind of measurement can be done at beamlines at synchrotrons around the world, as discussed in much more detail elsewhere (Egami and Billinge, 2012). In terms of data quality, high-energy x-rays produced by synchrotron facilities are optimal. The photon flux can be many orders of magnitude higher than what laboratory sources can produce, dramatically shortening the time needed to collect a full dataset. The higher-energy x-rays available penetrate samples better, allowing for measurement of data from samples with high Z elements, from thicker samples, and makes it possible to collect data from samples in more complex environments. Combined with the possibilities for fast measurements, this allows *in situ* or *operando* measurements, as we will discuss further in Chapter 14.

However, x-ray total scattering data can also be obtained from home-laboratory powder diffractometers. The measurements can take much longer, often 15–30 hours for one dataset compared to seconds or minutes at a synchrotron, though this time is coming down with the latest instrumentation. The beam energy is not tuneable, but various energies are available through the choice of anode material. A "standard" PXRD instrument with a copper anode will not give a small enough x-ray wavelength to cover a sufficient Q-range for PDF analysis, but other anode materials are available that allow PDFs to be obtained. Silver anodes can produce a Q_{max} comparable to synchrotron sources (Ag Kα_1 $\lambda = 0.5594$ Å, $Q_{max}(2\theta = 160°) = 22.12$ Å^{-1}). However, for weakly scattering materials, the statistics are often not sufficient to use the high-angle scattering data, and it can be beneficial to trade high energy for improved statistics, for example, using a molybdenum anode (Mo Kα_1 $\lambda = 0.7093$ Å, $Q_{max}(2\theta = 160°) = 17.45$ Å^{-1}). Powder diffraction instruments with Ag or Mo anodes are becoming more and more common to find in chemistry, geology, or physics laboratories. The calculation of the PDF from laboratory data requires different corrections than for synchrotron x-ray data, and dedicated software exists for this purpose.

For x-rays, the scattering power depends on the electron density of the sample, i.e. the species of atoms in the sample and the atomic density. The scattering power of an element is proportional to Z^2. Therefore, longer scattering measurements are generally needed to obtain high quality PDFs from, for example, samples containing light elements or from microporous samples. X-ray data from compounds with both light and heavy elements are dominated by the heavier elements, and it can be very difficult to discern structural information about oxygen in heavy metal oxides from x-ray PDFs.

2.6.2   PDFs from neutrons

So far, we have only discussed the use of x-rays for PDF analysis, but neutron PDF is an equally important technique. Luckily, we can easily extend the concepts described in this chapter to neutron scattering. There is a very important difference: neutrons scatter off the atomic nucleus, and neutron scattering intensity depends on the *neutron scattering length* b of the elements, which varies across the periodic table and among atomic isotopes in a way that appears random. This is in contrast to the x-ray atomic form factor f, which scales with atomic number Z. Neutrons can thus be very useful for distinguishing between atoms that are next to each other in the periodic table or for analysing structures containing light elements, such as lithium, where x-rays are challenged. Very different contrast properties can also be taken advantage of to get more information from lighter elements in compounds with heavy elements. Since neutrons also interact with magnetic moments, information about the magnetic structure in a material can be obtained, as we will see in Chapter 13.

Neutrons have a constant Q-dependence of the neutron scattering length b (unlike the x-ray form factor where f decreases with Q). This increases the high-Q signal relative to the noise compared to x-ray data. However, even at modern spallation neutron sources, the neutron flux is much lower than the x-ray flux at synchrotron sources, and neutron scattering measurements therefore take a much longer time and generally require larger amounts of sample than x-ray measurements. When preparing for neutron measurements, special considerations must furthermore be made for certain isotopes, e.g. hydrogen, that generates a significant amount of incoherent scattering making PDF analysis challenging due to a very large background signal.

Neutrons require specialized facilities for production, i.e. a nuclear reactor or a spallation source. Spallation sources operated using the time-of-flight method are generally preferred for PDF work due to the high flux of short-wavelength neutrons.

2.6.3   PDFs from electrons

More recently, the use of electrons for PDF analysis of nanostructured materials has also been developed: using standard transmission electron microscopes (TEMs), it is now possible to obtain data suitable for quantitatively reliable PDF analysis (Abeykoon *et al.*, 2012; Abeykoon *et al.*, 2015). In contrast to the special facilities required for neutrons or synchrotron x-rays, the electron beams generated from typical in-house TEMs can be suitable for obtaining PDF-quality scattering data. One of the benefits of electrons is that they interact much more strongly with the sample than x-rays or neutrons, resulting in a stronger scattering signal. This means that significantly less sample is needed to obtain a suitable measurement, and the electron beam can be tuned to a very small size for spatially resolved measurements. However, the strong interaction can also be a problem, as multiple scattering can take place, making electron PDF less reliable, and beam damage from the electron beam can also be significant. Nevertheless, "ePDF" has been shown very useful and is likely to become more widely used in the future (Abeykoon *et al.*, 2012; Abeykoon *et al.*, 2015)

In this book, we will deal mainly with x-ray data, but keep the possibilities in electron and neutron PDF in mind when considering your own scientific problems.

2.7   It's time to start modelling!

Small box modelling of PDFs will be the subject of the rest of this book. Through numerous examples, we will guide you through the steps of fitting a model to PDF data and answering scientific questions based on the results. We will start out with examples of well-ordered crystalline materials and work our way through nanoparticles, mixtures of phases, thin films, discrete clusters, and layered materials. While we hope this chapter has provided you with some of the background for understanding the relations between scattering data, atomic structure, and the PDF which should make you ready for rest of the book, we encourage you, when you get further into the method, to dive into the literature explaining scattering theory and PDF methods in much more detail (Billinge, 2008a; Egami and Billinge, 2012; Billinge, 2013). For now, start modelling!

3

PDF modelling of simple crystal structures: Bulk Ni and Pt nanoparticles

Kirsten M. Ø. Jensen and Simon J. L. Billinge

3.1   Introduction and overview

In this chapter, we will show the basic processes in PDF (pair distribution function) data modelling. At the same time, we will use the chapter to describe the layout of the book and how we intend you to use it, as all chapters will be built up in the same way. The Introduction and Overview section presents an introductory preamble to the problem in the chapter, including the main pedagogical goal of the problem. After reading the Overview and Introduction, you should be able to make a reasonable decision as to whether this problem is interesting enough to you that you want to try and solve it yourself.

The exercise in *this* chapter is to obtain structural information from the PDF of two samples with simple, well-ordered crystal structures: crystalline nickel and platinum nanoparticles. We will show you how atomic structure is represented in the PDF and introduce you to structural modelling using PDFGUI. We will show how to refine the structural parameters, such as lattice parameters and atomic displacement parameters (ADPs), and see how to include correlated motion of neighbouring atoms in your model. We will also learn how to determine a nanoparticle size from the PDF by taking into account the instrumental PDF damping.

The "solution" to the problem is at the end of the chapter. We recommend you only go there once you have found your own solution. There are some tips and tricks there from the experts, so even if you got a successful solution, it is worth a visit. It is, of course, also there if you get terribly stuck! Finally, we note that it is possible to do the same refinement using the more advanced DIFFPY-CMI software that we will be using in later chapters. We present a DIFFPY-CMI solution at the end of the chapter too. We recommend that you skip this, unless you are at the point that you are learning DIFFPY-CMI!

Atomic Pair Distribution Function Analysis. Kirsten M. Ø. Jensen and Simon J. L. Billinge, Oxford University Press.
© Simon J. L. Billinge and Kirsten M. Ø. Jensen (2023). DOI: 10.1093/oso/9780198885801.003.0003

The PDFs we are treating in the chapter were obtained from x-ray total scattering data – in a later chapter (Chapter 4), we will show you how this was done so you can do the same to your own data.

3.2 The question

In the "The Question" section we describe the scientific question that we were trying to answer with the PDF study. You should keep this in mind as you develop your strategy for working on the exercise, and indeed on all your own PDF studies.

In this chapter, the *scientific* goal may be "What is the average particle size of my Pt nanoparticles?", but really the main goal is a *pedagogical goal*: "What are the structural parameters for bulk Ni and Pt nanoparticles at room temperature, and how do we get them from PDF analysis?"

3.3 The result

In the "The Result" section we summarize the scientific result and give a reference to the paper or papers in the literature describing the result. Normally when you are doing your analysis, you do not know the result (you are trying to find it by doing the modelling!), though you may have some idea or hypothesis. However, since the goal here is to develop intuition and expertise in the analysis of PDFs, rather than actually to answer the scientific question, we thought it was a better idea to have the result in mind and to try and work towards it, rather than walking blindly. Nonetheless, SPOILER ALERT, don't read the "Results" section in chapters where you want to try and find the result yourself.

In the current example, the first result is the structural parameters of Ni, allowing us to present a model of the structure. More importantly, the analysis of the Ni PDF allows us to obtain experimental resolution parameters Q_{damp} and Q_{broad}, which characterize the instrument used to obtain the data. We then obtain structural parameters from Pt, and by using the Q_{damp} and Q_{broad} parameters obtained from the previous refinement, we are able to determine the size of the Pt nanoparticles.

3.4 The experiment

Since, sadly, you could not be at the experiment, in "The Experiment" section we report everything that went on at the experiment (well, everything that is relevant to the data analysis anyway!). In this case, the experiment was rather simple. We measured the x-ray total scattering signal from powders of bulk Ni and Pt nanoparticles enclosed in Kapton capillaries. We used the rapid acquisition *RA-PDF* setup [Chupas *et al.,* 2003] where a large-area 2D detector is pushed up close to the sample, and the sample is exposed to high-intensity, high-energy x-rays, as we also introduced in Chapter 2. More details may be found in Chapter 14 among other places [Egami and Billinge, 2012; Terban and Billinge, 2022; Chupas *et al.,* 2003]. We also collected data from an empty Kapton capillary so that we were able to correct for background scattering. Everything was done at room temperature. We used xPDFSUITE to obtain the PDF from our data file (you will learn how to do this in Chapter 4), which resulted in the files `Ni.gr` and `Pt-nanoparticles.gr`.

The experimental conditions are summarized in Table 3.1, and the files you need to download are listed in Table 3.2. Instructions for downloading files are given in Section 3.6.1. The `Ni.gr` and `Pt-nanoparticles.gr` files are text files containing two columns: r (usually in Å), and $G(r)$, the reduced pair distribution function. When working with PDFs, you will sometimes see `.gr` files with a third and possibly a fourth column. In the three-column format, the third column contains the uncertainties on $G(r)$, and when there are four columns, the third and the fourth contain the uncertainties on r and $G(r)$ values, respectively. All these files can be read into Python programs using the LOADDATA program in the DIFFPY.UTILS package that is available on Conda-Forge (see Apendix A for help installing software). As a last resort, if you are having trouble reading the files with your program, you can manually delete the header information, but this is not recommended for reasons of scientific reproducibility.

The `Ni.cif` and `Pt.cif` files are also text files, containing the structural information you need to set up your PDF refinements. As well as with a text editor, you can open `.cif` files in any structure visualization software, where you can see the atomic structure of the model.

Table 3.1 Experimental conditions for data collection.

Facility	NSLS-II
Beamline	XPD
Detector type	Perkin Elmer amorphous silicon 2D detector
Sample geometry	Powder in 1 mm ID Kapton capillary
Sample environment	Room temperature, ambient conditions
exposure time	60 s
Wavelength	0.1836 Å
Q_{max}	25 Å^{-1}

Table 3.2 Files for download.

`ni.gr`	Ni experimental PDF
`ni.cif`	Ni structure in a CIF-file
`Pt-nanopartices.gr`	Pt nanoparticles experimental PDF
`Pt.cif`	Crystalline Pt CIF

3.5 What next?

In this section we give a list of steps that you should follow if you are going to try the example for yourself. The boldest approach, only for the stout of heart, would be to try and figure out what to do just from a reading of the previous sections. In this section we give you a list of steps that we suggest you try and follow to help you to find your way to the answer. By design, the list is brief and does not tell you *how* to do the steps. We leave that to you to try and figure out on your own. However, if you get stuck on a step, you can peep ahead to Section 3.6, "Wait, What? How do I do that?". In that section, every step (or just about every one) outlined here has a description of how you might go about doing it. We hope that as you increase in experience, you can do more and more of the steps without referring to the "wait, what" crib.

Here are the steps of this example of modelling the Ni and Pt PDFs:

1. Download the PDFs (`.gr` files) and structure models (`.cif` files).
2. Sanity check (always a good idea): Plot the PDFs and consider similarities and differences.
3. Fit the Ni data in PDFGUI.
   (a) Load the `Ni.gr` data file in PDFGUI.
   (b) Plot the measured Ni $G(r)$ in PDFGUI.
   (c) Set up a starting model from the `Ni.cif` file or from scratch.
   (d) Save the project with a memorable name.
   (e) Calculate the PDF from the Ni model (without carrying out a refinement) as a sanity check of your starting model.
   (f) Set up and refine the structural and instrumental parameters in the fit until you get a good agreement between model and data:
       i. Refine the scale factor and the unit cell parameter in the data range from 0–20 Å^{-1}.
       ii. Refine the instrumental damping parameter.
       iii. Refine the parameters determining PDF peak shape.
       iv. Finish the refinement by including the full data range in the fit.
   (g) Plot the PDF with the fit model and a difference curve, and extract the refinement results.
4. Fit the Pt data in PDFGUI.
   (a) Load the Pt nanoparticle PDF file in PDFGUI.
   (b) Set up the starting model from the `Pt.cif` file or from scratch.
   (c) Set up the instrumental parameters using your results from the Ni refinement.
   (d) Refine the structural parameters until you get a good agreement between model and data.
   (e) Plot the PDF with the fit model and a difference curve and extract the refinement results.

3.6 Wait, what? How do I do that?

As we mentioned in Section 3.5 above, this section contains helpful hints about how to do each of the steps that were outlined there. Roughly speaking, there is a sub-section for each of the enumerated items in the "What next?" section that give further hints about how to proceed, if you need it.

3.6.1 Download the PDFs (.gr files) and structure files (.cif files)

All files you will use throughout the book are stored in an online repository at GitHub.com. GitHub is an online portal used by many programmers as a repository and file management system for their software code, but do not panic if you are not a programmer. It has some nice characteristics that we will make use of to maintain and distribute the database of files needed for these examples, and you can get them easily from GitHub in different ways, depending on how comfortable you are, or are not, with git, the distributed code repository software.

The steps to download the data and examples are the following. Download the data as a `zip` file and unpack it. At the time of writing the steps are:

1. Navigate to https://github.com/Billingegroup/pdfttp_data in a web browser
2. Click on the green button that says "Code"
3. Select "Download Zip"
4. Use your File browser to unpack the zip file into a set of folders on your file system.

There is one folder for each chapter. For example, for this chapter it will be in a folder called `Ch03NiModelling` wherever you unpacked it, but probably under a folder called `pdfttp_data`.

If you love git (as we do), then instead of downloading the bundle, you may want to clone the repository to your local computer filesystem. This produces a replica of the whole git repository onto your local computer. This is not recommended unless you are used to git but has the advantage that it is easy to get updates in the future.

3.6.2 Plot the PDFs and consider similarities and differences between them

Now we assume that you have the files successfully downloaded, unpacked, and sitting in a directory on your file system where you want to work. Use your favourite plotting program to make a plot of the two PDFs, contained in the two `.gr` files. They can also be easily plotted in PDFGUI or using the PLOTDATA functionality if you have PDFGETX3 installed (see Section 4.6.2). To plot them in PDFGUI, jump ahead to Section 3.6.6 to learn how to load them into the program. Then select the `Ni.gr` file in the Fit Tree window (left click it), and click the plot icon to plot it.

The files should be readily readable by many plotting programs as the `.gr` files are both text files with a number of lines of header information at the top, followed by the columns of data, as discussed in Section 3.4. The file header has information about how the data reduction was done, and these lines may be skipped if you are just importing the data for plotting. The PDFs are shown plotted in Figure 3.1.

When you have the two files plotted, first consider the Ni PDF. The PDF shows a number of well-defined, sharp peaks. This shows us that the sample is ordered, with well-defined interatomic distances. The position of the first peak is 2.49 Å, which tells us directly the interatomic distance between nearest neighbours in the Ni structure. The next peak is at the second neighbour distance, which is at 3.52 Å, and again that is evident in the plotted PDFs. All the other peaks also come from characteristic interatomic distances in the material.

We also notice that the first peak is bigger than the second peak. A number of things affect the size of the peak, such as the nature of the chemical species contributing to the peak, the nature of the scattering particle (x-rays or neutrons or electrons), and the multiplicity, or coordination number, of that coordination shell, as we saw in equation 2.7 for $R(r)$. Nickel is a particularly straightforward case because it is a chemical element, and so the only thing affecting the peak intensity (integrated area) of neighbouring peaks is the multiplicity, or coordination number, of those peaks; i.e. the number of atoms sitting at that distance from each other. The *fcc* structure (Fig. 3.1) is close-packed, and each nickel atom has 12 neighbours. The second neighbour peak comes from pairs of atoms on neighbouring faces of the unit cell. There are 8 such neighbours, which is fewer than 12, and so the second PDF peak is smaller than the first one.

We see sharp PDF peaks even at high-r values, showing that the Ni structure has long-range order, as expected for a crystalline material. However, you also see that the peak intensity dampens off with r. If we had a perfect instrument, with unlimited Q-resolution, we would see clear PDF peaks all the way to infinitely high r-values. The conditions for the measurements results in damping of the peaks at

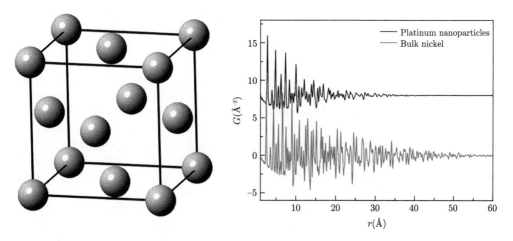

Figure 3.1 Ni structure (left) and PDFs from bulk nickel and platinum nanoparticles (right).

high-r, which we can refer to as "instrumental damping". We discuss what that means later in this chapter.

Now compare the Ni data to the PDF from the Pt nanoparticles. Just as for the Ni dataset, we see sharp, well-defined peaks, again illustrating that the structure in the sample is well-ordered. The position of the first peak is 2.78 Å, i.e. the Pt–Pt distance is slightly longer than the Ni–Ni distance, as expected from the atomic radii of the two elements. However, you can recognize the same pattern of peaks in the two PDFs. This is because both Pt and Ni take the face centred cubic, *fcc*, structure. The packing of the atoms in the two compounds is the same, giving comparable PDFs, but due to the difference in atom size, the peaks are shifted.

The main difference between the two datasets is an increased damping of PDF peak intensity in the nanoparticle dataset. This is due to the finite size of the nanoparticles, as the PDF will not show any interatomic distances beyond the size of the particle. We can estimate the particle size directly by observation (where the signal dies away in the plot, i.e. around 45 Å for the Pt nanoparticles) and more accurately through some simple data modelling as we describe below.

It will be discussed more later, but the particle "diameter" that is seen in the PDF is most closely related to the size of ordered domains in the sample and can be smaller than the physical size of the particle seen in small angle scattering measurements or transmission electron microscopy images.

3.6.3   Load the Ni.gr data file in PDFGUI

The software we will use for structure modelling is PDFGUI [Farrow *et al.*, 2007], which is our most user-friendly program written for PDF analysis. You can download this for free from diffpy.org, where you will also find installation instructions. Please also see Appendix A for more help with loading the software. PDFGUI comes with an extensive manual and tutorials. We will not reproduce the manual here (although we will see a lot of the features in the program as we go through the book) but rather introduce important steps in structural modelling so that you will understand the processes and how we get to the scientific results. It is recommended that you refer to the manual if you get lost in the buttons and specific procedures.

The workflow in PDFGUI is to create a "fit" by importing our experimental PDF and setting up a model for the structure that we will fit to the data. The "fit" involves "refining" the variable parameters in the model until a good agreement between the experimental and calculated PDFs is obtained. PDFGUI uses a crystallographic representation for the structural model, where the atomic structure of solids is described in terms of unit cells and fractional coordinates of atoms in the unit cell. As we will see in later chapters, this does not mean that we are limited to study crystalline materials with long range order, but for now, we will stick to crystallographic terminology. We will refine the structural parameters such as unit cell length and atomic displacement parameters, ADPs. However, it is not only the structure that is reflected in the data – the way the data were measured (the instrument) will also influence the PDF, which means that "instrumental parameters" must also be included in the model to fully describe the data. We will get through all of these parameters as we refine the Ni data step by step.

First, open PDFGUI. The layout (Figure 3.2) has a menu bar, a tool bar with functionalities that are used often, and four different panes: "Fit Tree", "Plot Control", "Output" and finally the main window, whose content will change as we set up the fit. First, right-click in the white area in "Fit Tree" to create your new fit. A fit icon will show up under the fit tree – it is from here that we will control the data and the model. Right click on the fit icon and choose "Insert dataset" in the menu that appears. This will allow you to open a file browser and navigate to the Ni.gr file. You can now open it, and you may notice that the file name appears in the fit tree with a small data-file icon showing up next to it. You can plot the file by first selecting (left-click) the data-file in the fit tree, then clicking the plot icon in the toolbar. The plot window should now appear, showing the PDF.

With the dataset selected in the fit tree (appears blue), you will see everything related to the dataset in the main PDFGUI pane. It has three tabs: `Configure`, `Constraints`, and `Results`. We'll get back to the last two below. In the `Configure` tab in the main PDFGUI window, a number of parameters have shown up describing the data you have just imported. These describe the conditions used in the experiments: scatter type (x-ray or neutron), the parameters used in the Fourier transform (PDF data range, Q_{max}), and the instrumental parameters that we will determine during the modelling (Q_{damp} and Q_{broad}). For now, leave these parameters at their default values.

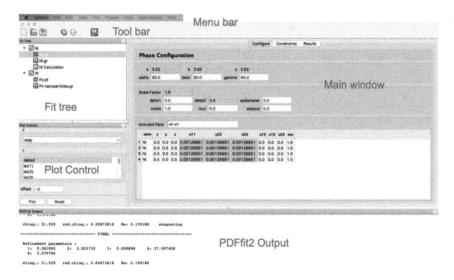

Figure 3.2 PDFGUI screenshot of the structure `Configure` tab after setting up the refinements.

3.6.4 Set up starting model from Ni CIF or from scratch

We will now set up the structure model. Ni has a well known structure, so we can find structural information in various databases, such as ICSD, COD, or CSD, or in the literature. In Chapter 6 we will also describe how structural models can be easily obtained from the structureMining [Yang *et al.*, 2020] app on the pdfitc.org [Yang *et al.*, 2021] website. The standard method of reporting structural data of crystalline materials is as a crystallographic information file (CIF), which contains information about the space group symmetry, unit cell, and atoms in the cell. The .cif file format can be read by most software for structural analysis, including PDFGUI. We will import our Ni structure model as a .cif, which you downloaded along with the data. Right-click on the fit icon, choose import phase, and choose "Load a structure from file". You can now use "Open", navigate to where you saved the files, and select the file you want to import into PDFGUI.

A structure icon will appear in the fit tree, which you can now click. All information about the structure model will show up in the main pane in the `Configure` tab, which you also see in Figure 3.2. Just as for the dataset, there are two other tabs associated with the structure, `Constraints` and `Results,` which we will get back to later. For now, click the `Configure` tab. You will see all the parameters describing the structure: unit cell, $(a, b, c, \alpha, \beta, \gamma)$, and atoms in the cell (fractional coordinates, atomic displacement parameters, occupancies). Ni has a cubic unit cell, so $a = b = c$, and $\alpha = \beta = \gamma = 90°$. There are four atoms in the unit cell as we see in the list of atoms. The four atoms are all symmetry equivalent and sit on special, high symmetry sites.

You can also construct a structure model from scratch. This is a useful option if you do not have access to a CIF but still know the relevant structural parameters (including all unit cell parameters, the space group of the structure, and the fractional coordinates of the atoms in the asymmetric unit). To do this, right-click on the fit icon, choose import phase, and choose "Create a structure from scratch" by clicking "New". Now, the phase configuration pane shows the parameters for a structure with a cubic unit cell with $a = b = c = 1$ Å containing no atoms. We should change this to a better initial guess for the Ni unit cell, so input 3.52 Å in the a, b, and c fields. We include atoms in the unit cell by inserting lines in the table in the lower part of the window. Right-click somewhere in the grey table header, and click on "Insert atoms". This lets you insert a number of rows in the table of atoms. The easiest option is to only initially insert the number of atoms in the asymmetric unit and then expand to all atomic positions by applying the symmetry operations defined by the space group. For Ni, the asymmetric unit is one atom sitting at (0,0,0), so insert one row and change the "elem" input from C to Ni. Make sure the (x,y,z) coordinates are (0,0,0), so that all other atomic positions can be generated through the symmetry operations defined by the space group. By right-clicking in the row with the atomic parameters, a menu will show up, where you can choose "Expand space group". You can now choose the space group for the Ni structure, $Fm\bar{3}m$, and after the expansion, all four positions show up.

If you have followed both instructions for inserting a structure (CIF or from scratch), you now have two structures in the fit tree – delete one of them, so that the fit tree just contains a dataset and one Ni structure. To do this, right-click on the structure-model that you want to delete in the fit tree.

3.6.5   Save the project with a memorable name

Having imported the experimental data and an initial model for the structural parameters, it is time to save the project. This is easily done in the `File` menu. This saves your work as a "project file" which has the name you choose for your project and has the extension `.ddp`. IT IS RECOMMENDED TO MAKE FREQUENT SAVES (CTRL-S is the keyboard shortcut for saving the project) because if the program hangs (and it sometimes happens), you can kill it and reopen the project without losing all your work. Every time you get to a place where you are happy with things, CTRL-S! If you get to a point where you have some more important intermediate or final result that you may want to return to, you can also select "save-as" and save an archival version of this version of the project, then reopen the working version of the project and continue playing.

3.6.6   Calculate the PDF from the Ni model without carrying out a refinement

Before we start the structure refinement, it is a good idea to calculate the PDF from the initial Ni model to check if it looks right. Right click on the Fit icon and choose "Insert Calculation" in the menu. The main pane will now show the configurations for the calculation. You can change the range to be from 0.5 Å to 20 Å and change Q_{max} to 25 Å^{-1} so that it is the same as for the experimental PDF. You now calculate a PDF from the Ni input structure with this configuration by selecting the calculation in the fit tree and clicking the blue gear icon in the toolbar. Select the calculation in the fit tree and click the plot icon in the toolbar to see your calculated PDF. Compare this to the experimental PDF – does it look right?

3.6.7   Refine the parameters in the fit until you get a good agreement between model and data

It is finally time to start refining the model to fit to the data! We do the refinement by choosing some of the parameters in the model to be variables. The refinement will then allow these variables to vary, until we obtain the best agreement possible between the calculated and experimental PDF. Generally, it is a good idea to start the refinement with only a few variables and then slowly add more as we get closer to a good fit. If all parameters are allowed to vary from the beginning, there is a good chance that the refinement will not converge, so we will do the refinement one step at a time.

We will initially do the refinement over a small r-range from 1–20 Å, so go to the `Configure` tab for the dataset and change the fit range to be 1–20 Å. Limiting the fit range in the beginning of a refinement speeds up the calculations and make it easier to obtain a good structural model. When our model fits well locally, we can extend the range to make sure the parameters we have refined describe the full dataset.

We set up the relationships between parameters and variables in the `Constraints` tabs. Each fit has two `Constraints` tabs, one for the structure, containing all structural parameters, and one for the dataset, containing all instrumental parameters. We start by going to the `Constraints` tab for the structure, shown in Figure 3.3. Here, you will recognize well-known parameter names such as the unit cell parameters (a, b, c, α, β, γ), as well as atomic fractional coordinates (x, y, z), and atomic displacement factors (U_{11}, U_{22}, U_{33}, etc.). The other parameters are the scale factor, controlling the overall intensity of the PDF; three parameters that can account for correlated atomic motion (δ_1, δ_2 and *sratio*); and a parameter related to structural coherence (*sp–diameter*), which we will come back to later.

Each of the parameters can be associated with variables (one single variable, or a mathematical expression of variables, as we will see in the following chapters), which are identified by numbers. We indicate that they are variables by preceding the number with an '@' sign – so @1 means variable 1; @23 means variable 23, etc. As we will see later, a powerful result of this construction is that we can easily create constraints between different parameters.

A good way to start the Ni modelling is to refine the scale factor of the PDF, which scales the overall intensity of the model PDF to the experimental data. Write "@1" in the scale factor variable input field (Figure 3.2). By doing this, you are telling PDFGUI that the value of the scale factor will be that of variable number 1. If you now click on the fit icon in the fit tree, you see a list of the refinement variables that are set to refine (Figure 3.4) – right now we just have the one, @1, but we will add more as we move on.

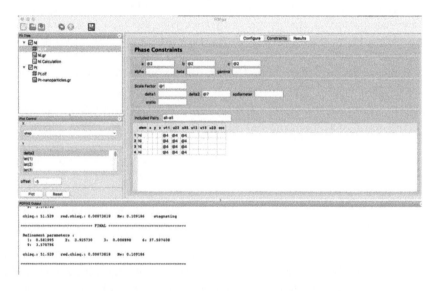

Figure 3.3 PDFGUI screenshot of the structure `Constraints` tab after setting up the refinements.

Figure 3.4 PDFGUI screenshot of the variable list after setting up the refinements.

Select the fit in the fit tree, and click the gear symbol in the toolbar – this starts a refinement. The scale factor is now refined from its initial value of 1.0 to fit better with the data intensity. If you again open a plot window by selecting the dataset and clicking the plot icon in the tool bar, the model PDF is seen along with the experimental data and difference curve. After we have refined the scale factor, we can see that there is some agreement between the two PDFs: their peak positions approximately agree. Since we are happy with the new value for the scale factor, we will accept the change: go to the variable list by clicking the fit icon, right click on the refined scale factor value (right hand column), and choose "Copy refined to initial" – the refined value will now be the starting value for the next refinement. Apart from seeing the results of the refinement in the refinement parameter list, you can also go the structure **Results** tab (remember, select the structure in the fit tree and then select the correct tab), where the refined value now shows up.

Next, we associate a variable with the unit cell parameters, which for the high symmetry Ni structure completely defines the positions of the PDF peaks. Since the unit cell is cubic, a, b, and c must all take the same value. Go to the structure **Constraints** tab, and write "@2" in the a, b, and c fields. This ensures that all three parameters are associated with the same variable. By doing a new refinement (by clicking the gear icon again), we will get a better fit of the PDF peak positions. When it is done, remember to copy the newly refined values of the scale and unit cell parameters to the initial values.

Now look at the data plot, again seeing the data PDF, the model PDF, and the difference curve. The intensity of the PDF peaks in the low-r region is too low, but it is too high at larger r values. The reason for this is the instrumental damping of the PDF, which was briefly described above. The damping can be included in our model by applying the Q_{damp} parameter. The Q_{damp} parameter attenuates peaks in the high-r region due to the finite Q-resolution of the instrument. You will find the Q_{damp} parameter under the dataset in the fit. As default, this is set to 0.001 Å^{-1} when starting a new refinement, but we will set it to 0.04 Å^{-1} as a better initial guess before we refine it – so first, go to the dataset `Configure` tab and change the value. Now go to the dataset `Constraints` tab and associate it with a variable, e.g. "@3". Do another refinement (blue gear icon in the toolbar), and remember to copy the refined values to initial when done. Our result from this fit is shown in Figure 3.5.

The peaks in the model PDFs (red line) are much narrower than in the data, which causes a large misfit. For Ni, the PDF peak shape comes mostly from thermal vibrations of atoms. We can take this into account by refining the atomic displacement parameters (ADPs) in our model. Go to the structure `Constraints` tab, and write "@4" in the U_{11}, U_{22}, and U_{33} fields of all four Ni atoms. This ensures isotropic vibrations, and make all four, symmetry-equivalent Ni atoms take the same ADP values as they are associated with the same variable. Adding thermal vibrations should drastically improve the fit of the PDF, as the PDF peaks get closer the right shape. Again, remember to copy the refined values of all the parameters to the "initial" column when the refinement is done.

At this point, you should be close to a very good fit – the red and blue lines almost completely overlap. However, there is still a slight misfit in the first peak in

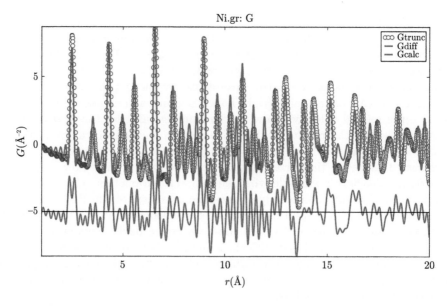

Figure 3.5 Screenshot from PDFGUI: Fit of Ni PDF after refining the scale factor, unit cell, and Q_{damp} parameters.

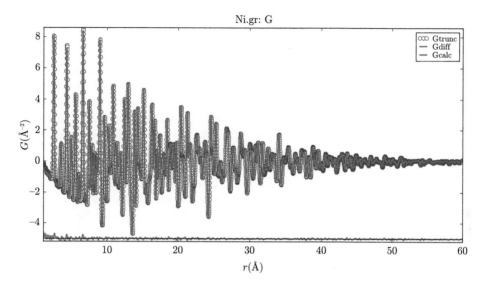

Figure 3.6 Screenshot from PDFGUI: Fit of Ni PDF after refining all relevant parameters.

the PDF, where the experimental PDF is slightly sharper than that of the model. This is because of correlated motion of neighbouring atoms. We will take that into account in our model by adding the parameter 'δ_2', which we will discuss in more detail below. A good value for the initial value of δ_2 is the position of the first PDF peak, here ~ 2.5 Å. Set the initial value, associate δ_2 with a variable, and refine!

The model now describes the peak positions, intensity, widths, and peak damping in r for our 1–20 Å fit range. Now when you have a good model, we will expand the fitting range to refine the entire PDF of 1–60 Å. This ensures that we get reliable values for the instrumental parameters, as these describe r-dependent effects and can only really be refined robustly when including the full PDF range. Change the fit range in the Configuration tab in the dataset and refine all parameters again to perform another refinement of the Ni data.

We can now add the final variable to the fit. Apart from thermal motion that we refined above, the PDF peak shape is also affected by the instrument – this can be taken into account by the Q_{broad} parameter. 'Q_{broad}' is an additional instrument resolution function correction parameter. It comes about mostly from the fact that diffraction peaks have different widths as a function of Q, which results in an r-dependent PDF peak broadening [Toby and Egami, 1992; Billinge, 1992]. This parameter is only required when data are fitted over a somewhat wide r-range. Go to the dataset **Configure** tab, give Q_{broad} an initial value of 0.01 Å^{-1}, associate Q_{broad} with variable @6 in the **Constraints** tab, and refine again.

3.6.8   Plot the PDF with the fit model and a difference curve and extract the refinement results

When opening the plot window, you should now see a good fit to the PDF data. If you want to save this plot as a image, you can click the "Save" icon in the plot tool bar. More importantly, if you want to export the data in the plot for further plotting, the PDF data and fit can be exported by going to the Data menu in the Menu bar and exporting the PDF fit as a fit data file (`.fgr`). This is a text file with all the relevant data: It contains a header giving information on the data collection and instrumental parameters, followed by data columns containing r-values, the experimental PDF, the calculated PDF, and the difference curve. This should be easy to import into your favourite plotting program for further plotting.

You can find all the refined parameters in the "Results" tab for the data set and structure.

3.6.9   Fit the Pt data in PDFGUI to determine the Pt structural parameters and crystallite size

Having finished our refinement of the Ni data, we can move on to data from the Pt nanoparticles. Our refinements will allow us to determine the structural parameters from Pt, just as we did for Ni. Furthermore, as we discussed above, we can use the PDF data from the nanoparticles to determine the crystallite size by modelling the size damping of the PDF peaks at high r. In order to do this, we need to know how much of the PDF damping arises from the instrument and how much comes from the particle size itself. Luckily, we can use the results from our Ni refinement to do this. The Pt and Ni data were collected at the same beamline with the exact same configuration (wavelength, detector-to-sample distance, beam size, capillary thickness, etc.). As the Ni sample is bulk and fully crystalline, the peak damping that we saw in the data and included in the model by fitting the Q_{damp} parameter fully describes the instrumental damping of the PDF peaks. By including the Q_{damp} (and Q_{broad}) in our analysis of the Pt data, without refining their values further, we know that any additional damping of the PDF peaks arise because of nanoparticle size. This allows us to determine the crystallite size of the Pt nanoparticles. When you do your own PDF studies of nanoparticles, always make sure to also collect data from a crystalline standard sample such as bulk Ni so that you can determine the instrumental parameters for your measurements.

We can now set up the Pt refinement. Start by adding a new fit to the fit tree by right-clicking in the white area and choose "New Fit". You can now import the data (`Pt-nanoparticles.gr`) and the structure model (`Pt.cif`) just as we did for the Ni data. Initially, set up the refinement in the dataset configuration tab to only include the range from 1–20 Å. We can then set up the instrumental parameters – copy the Q_{damp} and Q_{broad} values from the Ni results into the configuration pane for the Pt dataset. Here, it is important to use your Q_{damp} and Q_{broad} results from the refinements, including the full range in the Ni data – if only refining the smaller range from 0–20 Å, the refined Q_{damp} and Q_{broad} are not reliable.

We can now start the refinements as we did before. We start by refining the scale factor and the cubic unit cell parameter. Go to the `Constraints` tab for the structure, and write @1 in the variable field for the scale factor and @2 in the variable fields for *a, b,* and *c.* Having refined these and copied the refined parameters to the "initial" column, you can also include the atomic displacement parameters in the refinement by again assigning one variable (e.g. @4) to all the U_{11}, U_{22}, and U_{33} parameters. This should give you a refinement similar to that shown in Figure 3.7. The peak intensity does not match, as we have not included sample damping due to the finite size of the nanoparticles. We include this by using the '*sp–diameter*' parameter, which expresses the particle size in Å assuming a spherical shape. Before assigning a variable to it, we have to change the initial value to a better guess of the size. When set to 0 Å (the default value), no particle size damping is included in the model. We can set the initial guess to 50 Å – once this is done, assign a variable and run another refinement. Now, the PDF peaks should be much better described, except for the first peak at 2.79 Å, just as we saw for the Ni refinement. We need to take correlated motion of neighbouring atoms into account, so assign a variable to the δ_2 parameter after making a guess for the initial value, and refine again.

At this point, you should have quite good agreement between the data and the calculated model. However, to be able to get a reliable value for sp_{diameter} (the spherical particle size), we need to fit the data in a larger r-range, as the sp_{diameter} describes the r-dependent damping arising from the sample. Go to the dataset configuration tab and change the fit range to be 1–60 Å and refine again. Once this is done, you should have a good description of the PDF (you can see ours in Figure 3.8), so you can export your fit and find the final model parameters in the `Results` tabs.

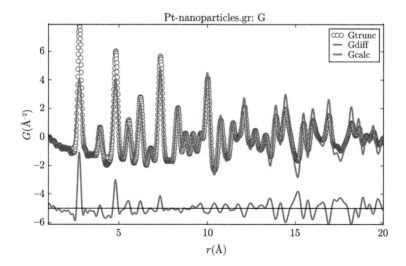

Figure 3.7 Screenshot from PDFGUI: Fit of Pt PDF after refining scale, *a,* and ADPs.

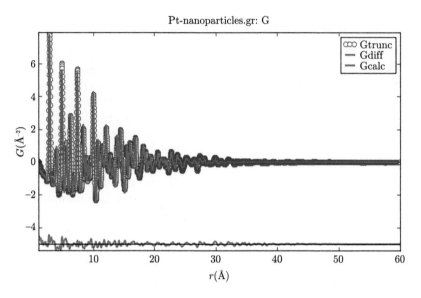

Figure 3.8 Screenshot from PDFGUI: Final fit of Pt nanoparticle PDF.

3.7 Problems

Here, we will give you some problems to consider based on your experience in the above sections:

1. Consider the parameters in the fit. Which parameters describe the structure, and which are characteristic of the instrument?
2. Why are so few parameters needed to describe the Ni structure?
3. What would happen to the experimental PDF if the dataset was measured at a higher temperature? Which parameters describe this effect?
4. Set the ADP values to zero and try and do a refinement. What happens? Why?
5. What is the physical origin for "correlated motion"?
6. Find the R_w values for the final fits of the two samples, expressing the quality of the fit. Which sample gives the best fit? Why?
7. Try to fix the sp_{diameter} to different values (between 10 and 100 Å), and see the effect in the PDF when you refine. What happens?
8. Which samples give reliable Q_{damp} values?

3.8 Solution

Our resulting fits are shown in Figures 3.6 and 3.8, and the refined parameters are listed in Tables 3.3 and 3.4.

Our data from the Ni powder could be described by refining only a few parameters: a scale factor describing the overall intensity of the PDF, the unit cell side

Table 3.3 Our refined values for the nickel fit.

Scale factor	0.455
Unit cell parameter (Å)	3.525
U_{iso} (Å^2)	0.0059
δ_2 (Å^2)	1.97
Q_{damp} (Å^{-1})	0.0453
Q_{broad} (Å^{-1})	0.0168

Table 3.4 Our refined values for the platinum fit.

Scale factor	0.582
Unit cell parameter (Å)	3.926
U_{iso} (Å^2)	0.0089
δ_2 (Å^2)	3.57
sp_{diameter} (Å)	37.6

length, one isotropic atomic displacement parameter (ADP), a parameter describing correlated motion, and two instrumental parameters, Q_{damp} and Q_{broad}. For the Pt nanoparticles, we only had to include one additional parameter, the sp_{diameter} expressing the size of the nanoparticles, and we set to fixed values (from the Ni refinement) but did not refine the Q_{damp} and Q_{broad} parameters.

This low number of structural parameters needed in the refinement is due to the high symmetry of the *fcc* structure, illustrated in Figure 3.1: The unit cell is cubic, so all the side lengths are equal and the angles in the cell are fixed at 90°. All four atoms in the unit cell are symmetry equivalent and sit on crystallographic special sites defined by the unit cell symmetry, and all sites are fully occupied. This leaves very few parameters that can be refined. In the following chapters, you will see how more complex structures will require refinement of more parameters, including fractional atomic positions, atomic occupancies, and parameters describing structural disorder.

During our refinement, we saw that the ADPs are important parameters when refining PDFs as these allow us to fit the PDF peak broadening. This is because the ADPs describe thermal motion – the higher the temperature, the more atomic movement and the broader the PDF peaks. The initial ADP values from the CIFs, 0.003 Å^{-2}, were too low to account for the peak broadening in our data taken at room temperature, but by allowing the value to increase, we were able to describe the data. Zero-value ADPs are unphysical, as these would express zero thermal movement of the atoms in the structure. In the PDF, zero-value ADPs would imply infinitely narrow functions, which cannot be calculated, and you will get an error message if you try. Sometimes, you will find CIFs in databases with all zero ADPs. If this is the case, you must change the ADPs to physical values before starting any calculations or refinements. Usually, values around 0.003–0.006 Å^{-2} are good initial guesses for the ADPs.

In both the Ni and Pt refinements, we assigned all relevant ADP values to one variable for all four atoms in the unit cell. You probably noticed that there are several parameters for each atom. This is because the ADPs can express atomic motion along all principal axes of the unit cell, *a, b,* and *c*. Most crystallography textbooks cover this in detail, so we refer you there for more information. The ADPs in a PDF model can be set up so that the atoms vibrate either isotropically or anisotropically by allowing one or more variables. For cubic structures like Ni and Pt, the high symmetry of the structure means that the ADPs must be isotropic, as there is no difference between the *a, b,* and *c* directions in the unit cell. We will learn more about how the space-group symmetry of the structure can be used to constrain the ADPs to symmetry-allowed models in the coming chapters. For now, we just note that even if the structure is not cubic, it is often sufficient to model the ADPs as isotropic. Introducing anisotropic atomic motion in your model significantly increases the number of parameters, and fitting anisotropic ADPs thus requires high-quality data to give reliable results. The models we use to describe the effect of thermal motion on the PDF are very similar to the Debye–Waller factors used in crystallography when modelling the effect of thermal motion on scattering intensity. If you are used to doing, e.g., Rietveld refinement of powder diffraction data, you may know "B-factors" better, but luckily, you can easily convert B to U or the other way around:

$$B = 8\pi^2 U \tag{3.1}$$

During the refinement we also added the δ_2 parameter to the fit, which sharpened the peak at the lowest r value to agree with the data. This effect comes from correlated motion of atoms. Using only ADPs, we assume that the atoms vibrate completely independently from each other. However, this is not the case for atoms that sit close together; their movement will be correlated, which sharpens the PDF peaks in the lowest r region [Jeong *et al.*, 1999]. You may have noticed that PDFGUI also gives the option of including parameters called 'δ_1' and 'sratio'. These parameters also account for correlated motion. The difference between 'δ_1' and 'δ_2' is very small (the parameters are described in detail in the PDFGUI manual), which means that they will strongly correlate if refined at the same time. When doing a refinement, you should thus choose which one to use, and fix the other to zero. The 'sratio' parameter can also be used to describe PDF peak sharpening but is not as commonly used as δ_1 and δ_2 – we refer the reader to the PDFGUI manual for further details.

We also saw the importance of the Q_{damp} parameter in the refinement of the Ni data. We described this as an intrinsic effect from the instrument related to the reciprocal-space resolution. The origin of the damping may be easier to understand if we think about the data in reciprocal-space, before the Fourier transformation to the PDF has been done: Consider the scattering pattern from a perfectly crystalline solid such as Ni. The Bragg peaks arising from the crystalline lattice should in theory be infinitely narrow lines. In reality, this is not the case. Many experimental factors, such as wavelength dispersion and axial divergence, lead to a broadening of the Bragg peaks, which get a finite width. For monochromatic, angular dispersive x-ray scattering experiments, as we have done here, the Bragg peak shape of a highly crystalline structure can usually be described by a Gaussian function with a certain width. When Fourier transforming x-ray scattering data to get a PDF, this corresponds to the Q_{damp} peak damping we see in the experimental PDF. If you are used to doing Rietveld refinements of PXRD data in 2θ or reciprocal-space, refining the Q_{damp} parameter corresponds to refining the instrumental peak shape parameters. At the end of the Ni refinement, we refined the Q_{broad} instrumental parameter. This also arises from limited Q-resolution, which induces an r-dependent broadening of the PDF peaks. In practice, this is only really seen when refining data over a wide r-range – i.e. 1–60 Å, as we used in the last refinement.

As discussed above, we were only able to find the size of the Pt nanoparticles because we had already determined the instrumental parameters. Whenever you collect PDF data for nanoparticles and you want to use the data to determine a size, it is therefore very important to also measure data from a crystalline "standard" sample as a calibration. Ni is a good choice, and other popular candidates are LaB_6, CeO_2, or Si. The most important point is that the standard should not give rise to any sample damping itself, so it should be a highly crystalline bulk powder with no strain in the structure. Such standards can be acquired from, e.g., the National Institute of Standards and Technology (NIST) and are often available at the beamline where you collect your data. Remember that you will only get reliable values for the instrumental parameters and particle sizes if you refine over a large data range.

As we went through the refinement steps, you could visually see the refinement getting better in the plot window. Apart from visual inspection (which is important!), the fit quality is also expressed through the R_w value, defined as:

$$R_w = \frac{\sum_i r_i (G_{obs}(r_i) - G_{calc}(r_i))^2}{\sum_i r_i G_{obs}^2} \tag{3.2}$$

You can find the R_w value in the PDFfit2 output window after each refinement and also in the "Results" tab appearing in the main pane when the fit is selected. For our final refinements, we got an R_w value of 0.0326, or 3.3% for the nickel data and 10.9% for the platinum data. The better the fit, the lower the R_w value, and it is therefore a good idea to follow the change in the R_w value as you are adding variables to the fit and comparing models. It is difficult to say when an R_w value is "good enough", as this will completely depend on the data, the structure, and the scientific problem as you will see as we progress though the book. Here, we are able to get a better fit for the Ni sample than the Pt nanoparticles. This is expected: The small Pt particles (ca. 4 nm) have a large surface area, where the atoms may not sit in a perfect *fcc* lattice, or the small size introduces defects in the structure. We furthermore model the particles as being monodisperse, while in reality, they are probably somewhat polydisperse. Introducing this in our model requires more advanced modelling tools or more complex models, as we will see in later chapters and as described in the literature [Banerjee *et al.*, 2018; Gamez *et al.*, 2017].

For now, we hope you have learned the basic steps in a PDF refinement and that you are ready to move on to scientific cases in the coming chapters! The following chapters build on the same ideas as we have gone though here, so remember to refer back here, or to the PDFGUI manual, if you are stuck on a parameter or two.

3.9   DIFFPY-CMI **Solution**

To make available a larger number of DIFFPY-CMI examples, we have included in the solutions section of the website a DIFFPY-CMI solution for every problem in every chapter. *You can skip this section on a first reading*, but it will be helpful to you when, in the future, you want to learn how to use DIFFPY-CMI .

A very good way to get started with DIFFPY-CMI is to work through the example, which has been developed as a Google Colab notebook. This can be run through your browser and does not require you to download and install any software. It provides all the code and data to run a beginning example. The notebook can be found with the solutions at https://github.com/Billingegroup/pdfttp_data. It also contains text that explains the different steps that are taking place in the code and is a very good way to play with DIFFPY-CMI and to get used to its basic structure. There are also videos that explain the use of diffpy products on the diffpy channel at YouTube. It contains a high-level summary of the structure of DIFFPY-CMI, and we reproduce the highlights here. Below we go through the actual example that is in the solutions folder that you downloaded

3.9.1   Running DIFFPY-CMI examples

To run the example in the Google Colab notebook, simply copy the notebook to your local Google account and then run it by hitting the "play" button on each cell. Some cells are hidden and can be opened by clicking on the right-facing triangles at the side of each section.

To run the examples that are provided in the "solutions" section of the download, you will need to download DIFFPY-CMI from diffpy.org and follow the installation instructions. Note that at the time of writing, DIFFPY-CMI will only run on Linux or Mac. It is possible to make it run on a Windows machine using the Windows Subsystem Linux (WSL). This is discussed somewhat on the diffpy-users Google Group (go to diffpy-users and type "cmi wsl" into the search window).

It is most likely that you will have installed DIFFPY-CMI in a conda virtual environment. In that case, to run the examples:

1. Navigate to the directory that contains the example script (or copy it to a directory where you want to work).
2. Edit the DIFFPY-CMI script so that it can find the data and cif files it needs, and knows where to put the outputs (see below for the Ni example).
3. Further edit the file to tweak what it does (see below).
4. Type `python <filename>.py` (where we always use the convention that quantities in angle brackets should be replaced with a particular value. In this case it would be `python fitBulkNi.py`).

That's it! In the future we will convert the Python files to Jupyter Notebooks, which may make them easier to follow.

3.9.2 The architecture of DIFFPY-CMI

Here we briefly describe the philosophy and architecture of DIFFPY-CMI to help you navigate the examples below. This section is easier to follow by running through the Google Colab example linked above. and we strongly recommend that you do that; but in case you don't have access to a computer/the internet, it is summarized here.

DIFFPY-CMI is designed as a flexible regression engine where you can do something as simple as fitting a straight line to some data or as complicated as a full-on multi-modal complex modelling problem with multiple datasets. The downside to this flexibility is that you have to build up the refinement script itself before doing the refinement. It would be like having a Rietveld refinement problem where you had to build the Rietveld refinement program before setting up and running your refinement. But a Rietveld program is only good if it does what you want to do. If what you want to do is not done by the program, that is the end of the road, but with DIFFPY-CMI it is just the beginning of the journey.

As a map for your journey, we describe the regression problem in general. First, there is generally some data and a physics or chemistry model that explains the data. We therefore need to have Python objects that hold the data (and enough metadata to explain what the data are and what they mean). We also need "calculators" that, given the physics/chemistry model, compute the measured signal. Finally, we need a mathematical model for the regression. The mathematical model requires a "cost function" that will be minimized (or maximized), a description of the variables that it can vary to do that optimization, and any constraints that we want to impose. Finally, we need a regressor that does that optimization. The regression engine knows how to update the variables in such a way as to optimize the function. As a concrete example, our signal may be a measured PDF, the physics model is a 3D arrangement of atoms, we have a calculator that, given a 3D arrangement of atoms can compute the PDF, the cost function is a chi-squared value for the agreement between the calculated and measured PDFs, and we may choose a least-squares optimizer like Levenberg–Marquardt, Newton's method, or simulated annealing to do the regression. If we felt the need to restrain the model so that it satisfied some bond-valence-sum (BVS) [Brown and Altermatt, 1985], or total energy, we could add a calculator that, given the arrangement of atoms computes the BVS or total energy, and we could add that to the cost function with some weight. If we know the structure is cubic, we could add a constraint that the lattice parameters $a = b = c$, for example, and so on. The only limitation is your imagination and the amount of time you have (and your coding ability!). Of course, developments you make can be shared back with the community through diffpy-users and the DIFFPY-CMI GitHub repository.

Here we briefly describe the Python objects and methods (functions) that you will interact with in the examples and which do this work.

Profile()   The Profile is an object to hold data and metadata. For example, in the Colab example we have a simulated dataset that is a linear line with noise. `Profile` is a general container for any profile. You will make a particular *instance* of it for each of your data profiles. In the Colab example, an instance called `noisy_linear` is created (instantiated) that contains our noisy linear data profile.

FitContribution()   This is roughly the equivalent in DIFFPY-CMI of each fit in the fit tree of PDFGUI. We use a `FitContribution` object to hold all the info about each contribution in the fit. In the Colab example we create a particular instance of FitContribution for fitting a linear model to the noisy linear data. We give this a short and memorable name (like the name of the fit in PDFGUI), and we give it our `noisy_linear` Profile object. Finally, we have to give it the physics/chemistry model, which could be a structure if we are fitting a PDF but is just a linear equation in the noisy_linear case.

FitRecipe()   This is the machinery that is the recipe for turning the physics models into mathematical models. The object to contain the complex fit is the `FitRecipe`, and we need to create a particular instance of this for each (single component) linear fit. We first create an empty recipe, give it a short and memorable name as before, and then we can add our fit contribution, or in general, as many contributions as we would like. We now do the important job of turning the physics/chemistry model into a mathematical model by mapping parameters (e.g. structural information like the lattice parameters) to variables (which the regressor will vary). This work was done by specifying the "@" variables in PDFGUI but is done with the `addVar()` method in the FitRecipe class. In general (non-linear least squares problems), we need to give an initial value for each variable, which PDFGUI takes from the `configuration` pane values or from the results of a previous refinement. In DIFFPY-CMI this is done with the `setValue()` method of the FitRecipe class. FitRecipe also has methods for setting constraints between variables and defining and calculating the cost function. With these the mathematical model is complete.

Optimization   `FitRecipe` is not in charge of the optimization of variables. It is only an interface to manage parameters and generate the residual. We need to use optimization tools outside DIFFPY-CMI, for example `scipy.optimize.least_squares`, but any regressor that you can make work can be used. After the regressor finishes its work, the refined values of the variables are stored in the FitRecipe object and can be accessed using the getValues() method. Calculated functions are also held in these objects allowing different things to be plotted, such as the calculated pattern, original data, and difference curve that we are used to seeing plotted.

3.9.3 The worked example in the Solution section

In this section in each chapter there will be some brief notes to help you understand the DIFFPY-CMI solution. The way to use this section is, first, to download the DIFFPY-CMI solutions for this chapter and open them in a text editor, or in your favourite Python IDE (see Section 16.4). You will read through the code in that example and look at the brief inline comments in that file. Since this is the first time that we are going through a DIFFPY-CMI solution, we present a slightly more expansive set of descriptions that refer by number to the numbered comments in the code that they expand upon.

3.9.4 `fitBulkNi.py`

This file presents a DIFFPY-CMI solution to the refinement of the Ni PDF with the Ni model that you did above. For this version of the file to work, we assume that you have downloaded the script in a directory called DIFFPY-CMI, with a parent directory called **solutions** and another directory called **data** located at the same directory level as **solutions**. The **data** directory should contain the `Ni.gr` file and the `Ni.cif` file. This is the arrangement in the downloaded files so should work "out of the box". If you have a different arrangement of your files, modify the code so that it can find the file locations.

Now, open up `fitBulkNi.py` in an IDE or text editor and start reading. When you see a line beginning with **#**, it is a comment. If it has a number, there may be a discussion here about it.

In this first introduction to DIFFPY-CMI, we will take a verbose approach to explaining what is happening. Subsequent examples will adopt a similar code work-flow, but we will be less verbose when explaining here in the text, focusing primarily on new aspects of the DIFFPY-CMI code. Of course, the Python source code will contain verbose commenting throughout.

#1: First we import all the packages that we need. Each import is actually a package with a name. The package has to have been installed in your conda environment so that Python can find it and import it. The file contains functions and classes and other things (we can call them objects), and these objects have names. If we see `import numpy as np`, this will import everything in the `numpy` package, and those things will be available to us by typing `np.<functionname>`. Sometimes the package name is too long, like `matplotlib.pyplot`, so we give it an alias, in this case `plt`, so a function in the package `matplotlib.pyplot`, for example `plot()`, will be available to us by typing `plt.plot()`. Sometimes we only want to import a few functions from the package, and we want them available to us with just their bare name, not `<pkgname>.<functionname>`. To do this we use the syntax `from`

<pkgname> import <functionname>, and then we can access them in the script as just <functionname>. Note that it is a Python convention to make variables that contain fixed global values all caps. Python variables that vary as the computation goes along are conventionally lowercase. Function names are also lowercase, but class names begin with an uppercase letter and are not all caps. Instances of classes are lowercase, so if you see something like me = Person(name = 'Simon'), you would understand it to mean that the variable me contains a particular instance of the Person class with the name "Simon". If the Person class has an attribute age, and it had been set or computed at some point, you could access it for me by typing me.age. If you are not sure what attributes and methods (methods are just like functions) there are, then in many IDEs you can ask by typing me. and then something like TAB or CTRL-RightArrow (check the documentation of your IDE) to see a list of everything that is available.

#2-4: Next there is a config section, which is pretty self-explanatory: it just defines some variables to point to things like the path to where the data are, DPATH, the name of the data-file, GR_NAME = 'Ni.gr', and a few other things. The FIT_ID (# 3) is a name for your fit, which is like the name you gave to the fit in the fit tree in PDFGUI . You could give these variables any name. They just hold on to these quantities and then will pass them to the refinement later. Though we refer to these as variables that have been assigned values, the values are never updated when the code runs, so they are really just global parameter definitions. In such cases it is traditional in Python to make the names all-caps, as mentioned previously. Everything will work fine if you don't follow this convention, but it makes your code more readable because other people reading your code will take the hint that these are just defining parameter values that will be used elsewhere.

#5: We can also set up some parameters that describe our experimental data. These were in the data config tab in PDFGUI.

#6-7: Below we will set up a refinement, and as usual it is a non-linear least-squares-type refinement, so the values of the refinable parameters (the variables) have to be initialized with some starting values. That is happening here.

#8: The overall structure of refinements in DIFFPY-CMI is that we make functions that do different things, then we make a special function, often called main(), that strings together these other predefined functions to run the refinement. In Python, the syntax for defining a function is def function_name():, and then the function definition is on the lines below this statement, indented by four spaces. Python uses the indentation to figure out when the function definition ends, so it is important that each statement in the function is indented by the right amount. The program will get confused and won't run if the indentation is wrong. Sometimes white-space can be very tricky too. The indentation may look right, but it is a **tab** and not four single spaces. Editors that are designed to write code, such as PYCHARM, know this and will automatically insert the right thing, but this can be a problem with

regular text editors. To be on the safe side, don't use the tab key (unless you are sure your editor is getting this right).

As described in Section 3.9.2, the heart of the refinement is the fit "recipe". You can build up your fit recipes by hand, but by way of a helper, we define a `make_recipe` function that we can reuse later and that helps us in principle to make new recipes with less editing and copy-pasting. Later, when we run `main()`, this function will execute and return back the fully formed recipe object.

The `make_recipe()` function is defined where it says `def make_recipe():`. Then later when we make use of it, we will give it a structure and a dataset, and it will know what to do with them to produce a recipe. We could make a new fit by running the same recipe on a different dataset, or the same data with a different model, and so on. Then we would reuse our recipe multiple times. We could even put it in a *for* loop and run the recipe on a large set of data, even thousands of datasets, in a highly automated fashion.

The `make_recipe` function takes two arguments, the full path to a file describing the structure, `cif_path`, and the data file we'd like to run, `dat_path`.

#9: First, when making our recipe, we load in the file representing the structure we want to use, located on disk at `cif_path`. There are many types of files for representing structures, and in this case we us the crystallographic information file (CIF) file format. We store the structure in a `structure` object. We also parse out the name of the space group, so that later we can use it to add meaningful constraints to our fit

#10: Next, we load in the experimental PDF data, located on disk at `dat_path`, and store it in a `Profile` object. The machinery for reading the information from disc and loading it into our program is handled by the `PDFParser` object (assuming we are loading a PDF). We can also get bits and bobs of information about our data by reading it from the file. For example, the `parseFile()` method handles loading $Q_{\min}$ and $Q_{\max}$ values from the `Ni.gr` file, if the information is present. In this case we are just fitting against one dataset, so we need only one instance of the `Profile`. The information loaded from the files by the parsers is then loaded into the our `Profile` object using the `loadParsedData` method. Other important user-definable parameters are then also loaded into the `Profile` object, such as the range of data and the mesh of points in r-space for the calculation. This is done using the `setCalculationRange` method.

#11: In this section we create and name an instance of the `PDFGenerator` class. This is the machinery that calculates a PDF given a structural model. In DIFFPY-CMI, it is possible to define other generators that will calculate given functions from a model, for example, the XAFS signal or BVS. In general, a fit could contain any number of generators to compute a wide variety of model signals. In this case, we luckily know we're just looking at the PDF signal from single phase nickel, so we just need one PDF generator.

#12: Next, we create and name an instance of the `FitContribution` class. In DIFFPY-CMI, fit contributions manage generators, allowing you, if you want, to combine or modify the signal computed from generators to build a more complex signal. We need to tell our newly created contribution about the generator we made previously, which we do using the contribution's `addProfileGenerator` method. As we mentioned, we can pass in additional generators, for example if we would like to consider a two-phase system (more on this in Chapter 5).

#13: In this code block we add the experimentally measured signal to the contribution using the `setProfile` method.

#14: As the PDF we compute may contain more than one contribution, or non-structural effects (such as finite domain size), we need to tell our fit exactly what to do with the PDF we calculate within the PDF generator we've created. We do this by providing an equation string to the `setEquation` method of the fit contribution. In this case, we define the equation string as `s1*G1`, where `G1` is the name of the PDF generator we created, and `s1` is a scale variable. If `setEquation` finds variables in the equation string it doesn't recognize, it will automatically add them to the fit contribution as new variables. Using this structure is a very flexible way of adding additional generators (i.e. multiple structural phases), experimental profiles, PDF characteristic functions (i.e. shape envelopes), and more.

#15: Finally we create a `FitRecipe` object which holds all the details of the fit, defined above. This organizes the whole fit, including constraints, restraints, all PDF generators, and fit contributions, and provides some useful methods for our refinement later on. We assign our fit contribution to the new fit recipe through its `addContribution` method.

#16: In this step we give initial values to the instrument parameters, Q_{damp} and Q_{broad}, and assign Q_{min} and Q_{max}, all of which are needed for the PDF generator to compute the PDF. Other parameters, such as lattice parameters, may be initialized automatically from their values in the CIF file. These can also be overwritten at this point if we don't like their initial values (for example, those pesky ADPs that appear as zeros in the CIF file sometimes).

Importantly, we are trying to get reasonable values for Q_{damp} and Q_{broad} here, such that they can be used subsequently for refining the Pt data. As Q_{min} and Q_{max} are generally fixed by the experiment, we use special methods to set their values. It's possible that the `PDFParser` method we used already found this information in the file header, but in case it didn't, we can set it explicitly again here. Better safe than sorry!

#17: Here we add the scale parameter we defined in the `FitContribution` using the `addVar` method. We also initialize the variable with a value defined in previous steps and tag it with a handy name. Tags are arbitrary but can be extremely useful later on if we want to free groups of parameters (for example, all atomic coordinate parameters) at once!

#18: In this step we make use of the space group name the parser found earlier. We can use the handy function `constrainAsSpaceGroup` to constrain the lattice and ADP parameters of our structure according to the relevant space group. First we establish the relevant parameters, then we can cycle through them and activate and tag them. In this case we must explicitly set the ADP parameters, because our CIF file had no ADP data.

#19: We need to tell our fit recipe about any other relevant parameters, otherwise we won't be able to vary them during our fitting. Here we add a correlated motion parameter (δ_2) and instrumental parameters Q_{damp} and Q_{broad} to the recipe. These parameters are part of the generator and are initialized with values as defined in earlier steps. We give them unique names of our choosing, and we don't forget to tag them with something meaningful.

#20: Finally, our fit recipe has everything it needs! At least one profile to hold the measured signal, a fit contribution with at least one profile, and one profile calculator (generator plus equation). We've added all the parameters we are interested in to the `FitRecipe` and applied all the relevant symmetry constraints. The `make_recipe` helper function then returns the built recipe back to whomever is lucky enough to call this function. Incidentally, this `make_recipe` helper function is not needed to make DIFFPY-CMI work. But it shows how you can make whatever helper functions you like and store them somewhere (for example, in GitHub). Then when you make scripts, you can import them and then run them, handing them as arguments whatever things they need to do their work and then receiving back the results of that work, such as a built recipe in this case.

#21: We might want to plot our fit later on, so here we define another helper function `plot_results`. This function takes two arguments, `recipe`, containing the recipe from which we'll plot a model and measured signal, and `fig_name`, which contains the full path and name of the file where we will write the plot.

We won't dive into detail on how this function is constructed, as it doesn't have much to do with DIFFPY-CMI, but feel free to peruse and read the comments!

#22: Now, we create the `main()` function we alluded to earlier. This is where all the pieces we've built previously will come together. In our case `main()` will take no arguments but will just run everything in the right order. Think of `main()` like a script.

#23: Here we call our `make_recipe` function, and we give it the full location of the structure file we want to load and the file containing our measured PDF. As discussed above, it will hand back a fully built `FitRecipe` object.

#24: Here we will run the actual fitting! Just as in PDFGUI, fit stability is improved if we free parameters in steps. So, we first fix all the parameters and then create a list of parameter names and tags we would like to free sequentially. We then loop over each entry in this list and free the parameter(s) associated with the item in the list. We make use of the `scipy.optimize` function `least_squares` to do the fitting, and it gets run at every iteration of our loop. `least_squares` takes at least two arguments, a function to be optimized and initial values for all the variables. Luckily in DIFFPY-CMI `recipe.residual` is just the function we are looking for that does this (it returns the chi-squared between the calculated PDF and the measured PDF, and as it happens in this case this is all we want to optimize), and `recipe.values` will give `least_squares` the initial values for all the variables.

#25: There is always a chance we will want to plot this fit again using an external plotting program later, so we make use of the our profile's `savetxt` method to write a text file containing the measured and model PDF signals.

#26: Of course, we would like to see the values of the fitted parameters, as well as other relevant information about the fit. We can compile the results with another built-in helper class, `FitResults`, and print them to the terminal with the method `printResults`.

#27: It's great to see these results printed to the terminal, but for the sake of posterity, we should also save them somewhere more permanent. We achieve this with the method `saveResults`.

#28: Finally we call our `plot_results` function, so that we can see our beautiful PDF fit and save it somewhere for use later.
   Congratulations, you have completed your first DIFFPY-CMI fit!

3.9.5   fitNPPt.py

This file presents a DIFFPY-CMI solution to the refinement of the nanocrystalline Pt PDF with the Pt model that you did above, where we use the values of Q_{damp} and Q_{broad} refined in our previous DIFFPY-CMI example but fix them, exactly as we did in the PDFGUI example.

   We again assume that you have downloaded the script in a directory called DIFFPY-CMI, with a parent directory called `solutions` and another directory called `data` located at the same directory level as `solutions`. The `data` directory should contain the `Pt-nanoparticles.gr` file and the `Pt.cif` file. Open up `fitNPPt.py` in an IDE or text editor and start reading.

   It's important to note that the previous example must be run completely before trying to run this example! The Python code is very similar to the previous example, so we will focus here on the key differences.

#7: In this example, we want to characterize the crystallite size of a platinum nanoparticle sample. To do this, we will use a characteristic function to damp the PDF, the sp_{diameter} parameter we discussed in the PDFGUI example. Here we choose a starting value for the crystallite size, in Å.

#8: In our last example, we refined values of Q_{damp} and Q_{broad} for our instrumental configuration. In this portion of the code, we read in these refined values from the output of the previously run example.

#15: We will try a characteristic function representing nanocrystals with a spherical morphology, by using the `sphericalCF` function. `sphericalCF` takes two arguments, the grid of points that we will calculate the function on (an array of r-values) and the nanocrystal diameter in Å, `psize`. We need to tell our fit contribution about this new function, which we do by using the contribution's `registerFunction` method. Here we can give our new function a name of our choosing, which we will use later when referring to it. Once we've registered the new function, we can define an equation as we did before for modifying the PDF computed by the generator. As before we multiply the output of the generator (`G1`) by a scale factor, but we also have to multiply it by the new function we registered (and which we called `f`; i.e. we define the equation string as `s1*G1*f`, where `G1` is the name of the PDF generator we created, `s1` is a scale variable, and `f` is our new spherical particle characteristic function.

#16-21: In these steps we set up the script to run the fit. Importantly, we do not add Q_{damp} and Q_{broad} as variables! These will remain fixed in this example, as we have determined them from the crystalline Ni fit.

#22-29: The remainder of our code proceeds identically as in the previous example on nickel. The only modification is to change which parameters we will fit over, as we now include our new crystal diameter parameter `psize` in the fitting loop.

4

Getting the PDF

Kirsten M. Ø. Jensen and Simon J. L. Billinge

4.1 Introduction and overview

In the previous chapter, you learned how to refine a structural model to a measured PDF. We had already obtained the PDF from synchrotron x-ray total scattering data. But in this chapter, you will do this yourself. Using either PDFGETX3 or XPDFSUITE you will learn how to go from a 1D scattering pattern to the PDF. We will show this using data from Ni as an example.

We will show all the main steps for both PDFGETX3 and XPDFSUITE, both introduced in Section 1.4.4.1. While they basically do the same thing (XPDFSUITE works as a GUI for PDFGETX3), using XPDFSUITE makes PDF life a lot easier and can significantly increase productivity in getting data analysis done. However, XPDFSUITE is commercial, while PDFGETX3 is freely available from diffpy.org. We will not reproduce the full manuals for the two programs here, so it may be a good idea to go through quick-start guides or tutorials to get used to the software.

4.2 The question

We will not really answer any scientific questions in this chapter, as we are still learning some of the fundamental steps in PDF analysis. However, we could pose the question as "How can I get a quantitatively accurate PDF from a 1D diffraction pattern?"

4.3 The result

In this chapter the result is an optimal PDF of bulk nickel. We provide examples of Ni PDFs that have been optimized for resolution and optimized for low-noise, and your job is to try and vary parameters until you get those results.

Atomic Pair Distribution Function Analysis. Kirsten M. Ø. Jensen and Simon J. L. Billinge, Oxford University Press.
© Simon J. L. Billinge and Kirsten M. Ø. Jensen (2023). DOI: 10.1093/oso/9780198885801.003.0004

4.4 The experiment

Total scattering data were measured from a Ni powder enclosed in a Kapton capillary. We also collected data from an empty Kapton capillary so that we can correct for background scattering. Everything was done at room temperature.

Note that .chi format files are text files that contain a header section and then data in two or three columns. In the two-column case, the first column is the momentum transfer Q or diffraction angle 2θ (depending on how the data were integrated), and the second column is the intensity. In the three-column format, the third column contains the estimated standard uncertainty on the intensity values. Occasionally you will encounter four-column format, in which case, most commonly, the third column is the standard error on the data in the first column, and the fourth column contains the standard error on the data in the second column.

Table 4.1 Experimental conditions for data collection.

Facility	NSLS-II
Beamline	XPD
Detector type	Perkin Elmer amorphous silicon 2D detector
Sample geometry	Powder in 1 mm ID Kapton capillary
Sample environment	Room temperature, ambient conditions
X-ray wavelength	0.1834 Å
Sample–detector distance	208.252 mm
Exposure time	60 s

Table 4.2 Files for download.

Filename	Note
Ni.chi	.chi format file of sample signal
kapton.chi	.chi format file of sample container signal
template.cfg	A template config file for PDFGETX3
ni-hires.gr	Target PDF optimized for resolution
ni-lownoise.gr	Target PDF optimized for low noise

4.5 What next?

1. Download the files.
2. Plot the two target PDF files to see what they look like.
3. Use PDFGETX3 or XPDFSUITE to generate PDFs:

(a) Set the configuration parameters to start the data reduction.

(b) Load the data files within PDFGETX3/xPDFSUITE to look at it, and generate plots of $I(Q)$, $F(Q)$ and $G(r)$.

(c) Dynamically play with the reduction parameters to see the effect on the PDF of Q-range and other parameters.

(d) When you are satisfied, try and set the parameters to make a PDF that is optimized for high resolution (sharp peaks).

(e) Save the data as a `.gr` file.

(f) Repeat the last three steps but optimize the PDF for low noise.

(g) Exit PDFGETX3/xPDFSUITE.

4. Make a plot of the target PDFs with the PDFs you obtained plotted on top for comparison.

4.6 Wait, what? How do I do that?

4.6.1 Download the files

Follow the instructions in Section 3.6.1.

4.6.2 Plot the two target PDF files to see what they look like

Plot those nickel files using your preferred plotting program. See Section 4.4 for help with `.gr` file formats. If you are using PDFGETX3, you can also use a plotting utility that comes with it. Open a command prompt or terminal window, navigate to the working directory, and type

```
> plotdata ni-hires.gr
```

to plot one, or

```
> plotdata ni-hires.gr ni-lownoise.gr
```

to plot them both on top of each other. PLOTDATA will plot all the input and output files for/from PDFGETX3 .

You can also use xPDFSUITE for simple plotting. The workflow in xPDFSUITE is to first use the file-system explorer to navigate to the working directory, then to select all files you want to work with into the active work area, then select the dataset (or sets) you want to plot and click on the 2D plot icon. You want to load in the two `.gr` files in your directory and plot them on top of each other. If you don't see the `.gr` files listed in xPDFSUITE it may be that the program is filtering for files with a different extension, such as `.chi`. There is a dropdown menu on the file browser to change what kinds of files the program filters for. If you change this to `all`, it will show all file types without filtering.

Now you know what your final PDFs are going to look like. The next step is to use our reciprocal-space data, i.e., the `.chi` files provided, to obtain plots like these.

4.6.3   Set configuration parameters and load the data

Both xPDFsuite and PDFGETX3 use only a few parameters to carry out the data reductions and analysis. Configuring the data reduction is slightly different in the two programs, and we give instructions for both here.

4.6.3.1   PDFGETX3

In PDFGETX3 the values of the configuration parameters are stored in a config file with an extension `.cfg`. To make it easier to get started, we have included a template, `template.cfg`, with the download files, so it should already be there in your directory. The PDFGETX3 config file is a plain text file, and you will use a text editor to edit it, so open `template.cfg` in your favourite text editor, for example Notepad on a Windows computer. It is a good idea to save it with a different name so it does not get overwritten by another template file or muddled up in some other way. Maybe "save-as" `my-hires.cfg` since you will be working on parameters for a high-resolution Ni PDF shortly. Then explore around inside the file to see how it is laid out, and see if you can figure out how to edit the relevant parameters in the file appropriately to get started. The first input line in the config file is the data format, where you can choose between diffraction angle (2θ) or momentum transfer Q (in Å^{-1} or nm^{-1}). If you open the data file (`Ni.chi`) in another text editor, you can see that the format in this case is Q in Å^{-1}, so give the input "QA" in "dataformat." Go through the rest of the config file, and fill out the other relevant input lines: Give the background file name (given above in Table 4.4) and the sample composition (Ni). Keep the starting values for *bgscale, rpoly,* $Q_{\text{max–inst}}$, Q_{min}, and Q_{max} at default values – we will get to them in Section 4.6.4.

There are many more options for inputs in the config file that can easily be applied when you are more familiar with the PDF and the software. We will not need them now, but for more information, read through the PDFGETX3 manual that can be found on the diffpy.org website.

You now want to run PDFGETX3 with the `Ni.chi` data file and your configuration file. You will have to again work in a terminal/command-line window. Reuse the one you used earlier, or open a new one and navigate to the working directory, then type:

> PDFGETX3 `Ni.chi` —config=`'myfile.cfg'`

Remember to replace 'myfile.cfg' with your actual filename. If everything is set up correctly, a window with plots of $I(Q)$, $F(Q)$, and $G(r)$ should now open. If you get an error message, check the config file again, and make sure that all necessary inputs are given. You have now generated the PDF! The next step is to optimize it with appropriate parameters as we discuss below.

4.6.3.2   xPDFsuite

If you are using xPDFsuite, all parameters involved in the PDF calculation appear in the GUI interface, and you can update them in there. Open the program from the desktop. In the right-hand pane under the **Basic** tab, you can give inputs on the data and sample you will be working with. The first input to give is the data format, where you can choose between diffraction angle (2θ) or momentum transfer Q (in Å^{-1} or nm^{-1}). If you open the data file (Ni.chi) in a text editor, you can see that the format in this case is Q in Å^{-1}, so choose Q in Å^{-1} in "dataformat." Next you can load the background file by navigating to the file obtained from measurements of only the sample container, with filename given in Table 4.4. The composition of the sample is simply Ni. You can also change the r-grid that the PDF will be calculated on, but for now, you can stick to the default values.

Now use the file browser to load the Ni.chi file from your folder. Double click on the .chi file. Now a window should appear with the PDF. Click "xrd," and $F(Q)$ in the top right-hand corner as well, to see both the raw data $I(Q)$ and the reduced total scattering structure function $F(Q)$ along with the PDF.

4.6.4   Play with the reduction parameters

XPDFSUITE and PDFGETX3 use the same parameters in the data reduction. We now want to play with them and see the effect on the resulting PDF, and we give instructions to both programs while describing the effect of the parameters.

In PDFGETX3 the parameters are adjusted by typing

```
> tuneconfig()
```

in the terminal from which PDFGETX3 is running. Now, a window with slide bars shows up that can be used to adjust the configuration parameters that we will describe below.

In XPDFSUITE the parameters are tuned directly through the GUI in the PDF tab in the right-hand panel, where you also see slide bars for each of the parameters.

4.6.4.1   Background scale

We first want to adjust the background scale. In our case, the background is the scattering signal measured for an empty Kapton capillary. The Kapton pattern was measured for twice as long as the Ni pattern, and the x-ray flux may have been slightly different. We therefore need to adjust the background scale so that it matches the Ni pattern. If considering the data in reciprocal-space (the top window in the plot panel), the Kapton signal has a bump in the low Q-region, which is seen both in the empty Kapton data and the Ni sample data. This bump is the scattered intensity from the amorphous Kapton tube, and we want to remove this signal from our Ni data before we obtain the PDF. Since we know that the Ni sample is completely crystalline, we do not expect its structure to give rise to any scattering intensity in that region, so we can adjust the background scale so that the Ni and Kapton data match up in the bump. Note that if you are working with samples that do show diffuse scattering themselves, you have to be more careful with determining the value for the background scale, as we will discuss more in Chapter 10.

Background scale tuning is easily done in XPDFSUITE, where you can directly plot the background data along with the sample data by checking the "Plot background" box in the plot window. You can now adjust the background scale with the slider until the two bumps are at the same intensity.

When using the default `tuneconfig()` function in PDFGETX3, you will not see the raw scattering data or the background data directly in your plot, but we can change the settings for tuneconfig. Close your current plot window, and in your terminal, where you are now in the "interactive" PDFGETX3 mode, you can type:

```
> t2 = pdfgetter.getTransformation(2)
> tuneconfig([t2, 'fq','gr'])
```

This should open new plot and `tuneconfig` windows, where you will see the measured scattered intensity from the sample, the measured background intensity, and the difference between the two plotted. This should make it easier to find a good value for the background scale, as you can now use the slide bar to find a value where the scattered intensity and the background intensity line up. We refer you to the PDFGETX3 manual to learn about `pdfgetter` and `getTransformation`.

4.6.4.2 $Q_{max\text{-}inst}$ and rpoly

The next parameters to adjust are $Q_{max\text{-}inst}$ and *rpoly*. Both these parameters have to do with the correction algorithm used before performing the Fourier transform of the data. XPDFSUITE and PDFGETX3 use an *ad hoc* approach to PDF reduction instead of doing explicit corrections for, for example, incoherent Compton scattering and fluorescence. This has been shown to give equivalent results to carrying out the full corrections in general (Juhás *et al.*, 2013), except for the overall scale factor of the data. The uncertain scale factor doesn't affect the *relative* heights of PDF peaks or the ability to extract accurate structural parameters from modelling.

The *ad hoc* approach is described in detail in Juhás *et al.* (2013), and it is important that you understand this process when you get further into PDF analysis. For now, just note that the program will determine a polynomial function that approximates all contributions to the measured scattering intensity that *are not* the signal (the coherent, elastic scattering). The polynomial will then be subtracted from the data, leaving only the contributions that are interesting to us.

For the program to do this, two inputs are needed:

1. *rpoly*, which is roughly the lower limit of r (in Å) beyond which we have reliable PDF peaks (it usually takes a value somewhat below the distance of the nearest neighbour distance in the material). This variable ensures that the polynomial function won't be flexible enough to remove actual structural signals by mistake.

2. $Q_{max\text{-}inst}$, which defines the range of data that the polynomial function will be fitted over.

Practically, we set $Q_{max\text{-}inst}$ to cover the maximum momentum transfer, Q, where the $F(Q)$ data are "well behaved," i.e. the region where $F(Q)$ doesn't have large non-physical fluctuations. These can get very large in the high-Q region of $F(Q)$, and we don't want the polynomial to have to fit crazy non-physical oscillations, so we throw them away. Our normal workflow is to plot the data over the *entire* measured range, so we can see all that badness at the edges of spectrum. Then we select a point on the curve that is the highest point in Q before all the badness and select this value for $Q_{\text{max-inst}}$. In XPDFSUITE this can be done with a slider or by hovering the mouse over the plot of $F(Q)$ and reading off the Q value and typing it into the relevant box in the GUI. For PDFGETX3, determine the best value from the plot, and type it into the config file.

The reason for having $Q_{\text{max-inst}}$ is that it makes the fits more stable if the data can be fitted over a range that is longer, in general, than the range of Q that will be Fourier transformed. This is why we define two distinct Q_{max} values. Thus, $Q_{\text{max-inst}}$ is an instrument parameter, which is not related to your sample but is given by the geometry of the measurement, the quality of your detector corrections, how you acquired the data, and so on.

The Ni data were measured using a square 2D detector in the *RA-PDF* setup and have subsequently been integrated to yield the 1D pattern provided here. We describe the integration process in Appendix B. From the intensity file, we see that the highest Q value with data is 31.26 Å^{-1}. However, the data obtained at the highest scattering angles, corresponding to the highest values of Q, come from only a few pixels in the corner of the square detector, and we will therefore not include these in the physical Q-range. Instead, we will use the highest value of Q where the detector covers the full azimuthal range of the diffraction pattern. For the current Ni data, we set $Q_{\text{max-inst}}$ to 29.5 Å^{-1}. The *rpoly* parameter can often be kept at the default value of 0.9 Å, so do not change that for now. This parameter is more important when the signal is weak, as we discuss further in Chapter 10.

4.6.4.3 Q_{min} and Q_{max}

Q_{min} and Q_{max} define the data interval that is used in the Fourier transform. Q_{min} should generally be set as low as possible; however, the beamstop (used to protect the detector from the direct x-ray beam) defines the minimum usable Q-value. It is advised to zoom in to the low-Q region of the intensity data and to see where the data start dropping down rapidly. A good estimate for Q_{min} is to place it on the high-Q side of this drop-off point. In our example data, zooming in to the low-Q region, you can see that below 0.8 Å^{-1}, the data are very noisy, so we cut it off there.

Q_{max} is the upper data limit used in the Fourier transform. By extending Q_{max} to high values, we minimize the effect of termination ripples in the PDF and get higher resolution in r, meaning that we can better distinguish neighbouring PDF peaks. However, at high-Q values, the data will also often be very noisy due to the decreasing x-ray scattering power f at high momentum transfers. We do not want to include the noisiest part of the data in our PDF, so most often the Q_{max} chosen is a compromise between resolution and noise. Use the slide bars to see what happens when you change Q_{max}. As you will see, the Ni data used here are of high quality, so even at the highest values of Q, there is not much noise. This may be different when you start working on your own data from other samples, and in Chapter 10, we will treat data with a lot more noise, where this compromise becomes more important.

4.6.5 Save the data as a .gr file

We first want to save a PDF with high resolution, so choose a suitable value for Q_{max}. In XPDFSUITE, you simply click on the gear icon in the top right-hand corner, select "Save data files," chose an appropriate file name, and click the files you want to save: iq, sq, fq, or gr.

In PDFGETX3, files were actually already saved when you first ran the `pdfgetx` command. However, these were processed with the initial configuration values. Now that you have changed them, you have to make the program overwrite the old files. You can do that by running:

```
> processFiles()
```

in the terminal running PDFGETX3. This takes the current configuration and over-writes the old files, if "force" is set to "yes" in the config file. You can change the outputs that you want `processFiles()` to save by updating the config file under the "outputtype" field. More details can be found in the PDFGETX3 documentation.

4.6.6 Repeat the last three steps but optimize the PDF for low noise

Instead of optimizing for high resolution, now try to limit the Q-range by choosing a lower value for Q_{max}, e.g. 15 Å^{-1}. Save your new PDF, as described above.

4.6.7 Exit PDFGETX3

Now close PDFGETX3 by typing

```
> exit()
```

If you are using XPDFSUITE, just keep it open – we will use it for plotting.

4.6.8 Make a plot of the target PDFs with the PDFs you obtained

Now compare your two PDFs with each other and with the ones we provided – plot the data however you like. If you are using XPDFSUITE, you can plot the data there, as described above. If you are a PDFGETX3 user, you can again use the PLOTDATA program from the command line by typing the following with appropriate file names:

```
> plotdata file1.gr file2.gr file3.gr file4.gr
```

4.7 Results

You have already seen the results – if you got good PDFs similar to the ones provided, you have succeeded!

4.8   Problems

Here, we will give you some problems to consider based on your experience in the above sections:

1. Consider your two PDFs. What gives the difference in peak width between the two?
2. How does noise in the measured data manifest itself in the PDF?
3. What factors affect the optimal value to choose for the Q_{max} parameter during data reduction?
4. What happens if you set the *rpoly* value to zero?
5. What happens if you do not subtract the Kapton background from your data before obtaining your PDF?

4.9   Solution

In Figure 4.1 we show the PDFs, i.e. the $G(r)$ functions, of Ni that were distributed with the data. The high resolution PDF is in blue and the low resolution one is in red. The two PDFs shown in Figure 4.1 were obtained from the same dataset and therefore from Ni at the same temperature. Nonetheless, careful inspection shows that the PDF peaks have different widths! In Chapter 3, we told you that the width of the PDF peak depends on the atomic movement, which we described by ADPs. However, now we see that there is also a data reduction effect on the peak width. Because of the Fourier relationship between the measured data and the PDF, the real-space resolution, i.e. the r resolution, is related not to the Q-resolution in reciprocal space but to the *range of data* in reciprocal space. The red PDF is obtained from the same initial dataset but was processed with a lower Q_{max}, resulting in slightly broader PDF peaks. For this reason, you should *only* directly compare PDFs with each other when they have been processed with the same Q_{max}. Since Q_{max} is a known quantity, its value is given to PDF modelling programs which correct for the resolution effects so that accurate thermal/disorder distribution widths may be obtained from the modelling, regardless of the Q_{max} chosen.

Having seen the advantage of a high Q_{max} (sharper peaks and a higher resolution measurement), why would we not always take data over the widest measured range possible? Well, we do tend to do that; however, another factor is that by extending the Q_{max} higher and higher, we introduce more and more measurement noise into the PDF because the signal-to-noise ratio falls off rapidly with increasing Q in the measurement. This is evident in the $F(Q)$ function plotted in Figure 4.2 where the corresponding $I(Q)$ is also plotted. In the high-Q region the signal becomes a little more noisy. This effect is greatly exaggerated in weakly scattering samples or for *in situ* data, where fast data collection is often important. At the cost of some real-space resolution, one can reduce the noise in the data by lowering Q_{max} somewhat. Where to place Q_{max} then becomes a compromise between these two ideals, and the value chosen will depend to some extent on the scientific question under study in a particular situation. We will revisit this point in Chapter 10. If answering the

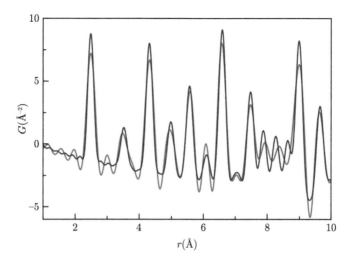

Figure 4.1 Our two PDFs: Red shows the PDF obtained with lowest Q_{max}, blue the PDF obtained with higest Q_{max}.

question depends sensitively on seeing small features close to the baseline that are not overlapped with other peaks in the PDF, then it is better to err on the side of a lower Q_{max}. On the other hand, if it is important to separate signals from two bond-lengths that are very close to each other in distance, then a higher Q_{max} is preferred. If the data become too noisy at the required Q_{max}, then it is necessary to measure for longer, or with a more powerful beam, to get adequate counting statistics at the desired Q_{max}.

We already discussed a bit the role of $Q_{\mathrm{max\text{-}inst}}$ and *rpoly* when obtaining the PDF, as these are used in the *ad hoc* data correction applied in xPDFsuite and PDFgetX3. If you set the value of *rpoly* to 0 Å, you are basically telling the program not to do any corrections to the data, and you will see that non-physical features will show up in the low r range. For a more weakly scattering sample with less clear Bragg peaks, these effects are much worse and will completely mess up your PDF. As mentioned above, r-poly is the lower limit of r (in Å) beyond which the sample has structural PDF peaks after the *ad hoc* corrections and Fourier transform, so you should also not use a high value. The default value of 0.9 Å is usually a good starting point, as this is shorter than most chemical bonds.

If considering the data from the Ni sample, you can see that the Ni Bragg peaks are much, much more intense than the scattering signal from the Kapton capillary. For this reason, you may not see much difference in the PDF with and without background subtraction, as the contribution from the Kapton tube is very small. This is not the case for weakly scattering samples, where the background signal can contribute just as much or even more than the sample itself. For many experiments, the background signal may even contribute the large majority of the signal – this could be the case, for example, in *in situ* experiments of particles and clusters in

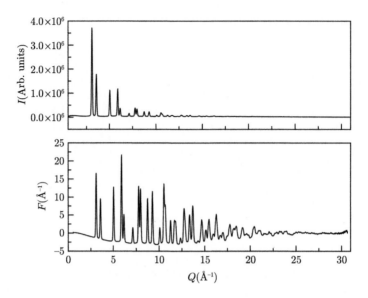

Figure 4.2 Ni $I(Q)$ (top) and $F(Q)$ (bottom).

solution, or in PDF measurements of thin films on a substrate, as we will see in Chapter 10. Especially for such data, the background subtraction is a crucial step in the data reduction.

With this chapter, you now know how to generate PDFs from integrated total scattering data. We will now move on to more advanced PDF modelling.

5

Quantification of sample phase composition: Physical mixtures of Si and Ni

Emil S. Bozin and Simon J. L. Billinge

5.1 Introduction and overview

It is common that a material system of interest is a mixture of structural phases. At times this could be a nuisance, for example due to the presence of impurity phases that have formed in the synthesis or traces of synthesis starting materials. Another and more interesting possibility is that the phase coexistence is intrinsic to the material and as such could be relevant for its properties. Irrespective of the origin of the multiphase nature of the system at hand, it is important to characterise as thoroughly and as reliably as possible the mixed phase character of samples. In conventional powder diffraction, this matter is handled by "quantitative phase analysis" (QPA) (Madsen *et al.* 2019; Dinnebier and Billinge, 2008) by using multi-phase Rietveld refinements (Toby, 2019; Dinnebier and Billinge, 2008). The starting point of this process is the phase identification, a step that is done prior to quantification to account for all the observed reflections in the measured diffraction pattern. Once all ingredient phases are known, the next step is a simultaneous Rietveld refinement of these ingredient phases against the data, which results in quantification of the relative abundance of these phases within the material studied and, of course, the structural parameters of each individual phase. Phase analysis can be carried out within the PDF framework as well, by using a multi-phase refinement. Towards the end of the chapter, in Sections 5.7 and 5.8, you will be prompted to think about the complementarity of the Rietveld and PDF approaches to the multi-phase scenario, about some common situations where these approaches could be handy, and about the applicability limitations and typical issues that one faces when dealing with a multi-phase situation, allowing you to grasp the complexity of the multi-phase problem.

Atomic Pair Distribution Function Analysis. Emil S. Bozin and Simon J. L. Billinge, Oxford University Press.
© Simon J. L. Billinge and Kirsten M. Ø. Jensen (2023). DOI: 10.1093/oso/9780198885801.003.0005

5.2 The question

In this chapter we do not solve an important scientific problem but describe a model system: a refinement of PDF data from a sample consisting of a mixture of Si and Ni powders. To prepare the sample, Ni and Si powders (micron-sized grains) were carefully mechanically mixed in ethanol using mortar and pestle. The PDF that we will model was obtained from x-ray total scattering data from the sample at room temperature. Data were also collected on ingredient materials, pure Si and Ni powders, to provide references for the ingredient phases. The primary scientific question is, "What are the phase fractions for the sample composed of admixed bulk Si and Ni powders at room temperature, and how do we get them from the PDF analysis?". Another important question could be "What are the lattice and atomic displacement parameters of the ingredient phases, and how do they compare to those obtained from the reference data?".

5.3 The result

After doing the refinement, we see that it is possible to obtain phase fractions that agree with the known weight ratio of the two phases in the sample. We also see that structural parameters can be obtained for both phases in the mixture, although the refined parameters from the minority phase in the mixture show slightly larger deviations from the results obtained from the pure phases.

5.4 The experiment

The experiment was carried out using the standard *RA-PDF* setup at 300 K on three samples: Si powder, Ni powder, and synthetic powder comprised of admixed Si and Ni powders in predetermined weight ratio.

Table 5.1 Experimental conditions for data collection.

Facility	APS at Argonne National Laboratory
Beamline	6-ID-D
Energy	86.84 keV
Detector type	Perkin Elmer amorphous silicon 2D detector
Sample geometry	Powder in 1 mm ID Kapton capillary
Sample environment	Room temperature, ambient conditions
Total exposure time	60 s
Q_{max}	27.0 Å^{-1}
Q_{damp}	0.05 Å^{-1}

Files to download:

Table 5.2 Files for download.

Filename	Note
si.iq	Intensity versus Q file of silicon reference
ni.iq	Intensity versus Q file of nickel reference
sini.iq	Intensity versus Q file of silicon-nickel mixture
si.gr	PDF, $G(r)$, versus r file of silicon reference
ni.gr	PDF, $G(r)$, versus r file of nickel reference
sini.gr	PDF, $G(r)$, versus r file of silicon-nickel mixture

For this example we provide six files. For each of the three samples there is an .iq file containing intensity versus Q, $I(Q)$, which is a two column ascii file. The first column is the momentum transfer, Q, in units of Å^{-1}, and the second column is the measured intensity, $I(Q)$. For each sample there is a pair distribution function versus interatomic distance r file, $G(r)$, which is a two column ascii file. The first column is the interatomic distance, r, and the second column is the measured PDF, $G(r)$.

5.5 What next?

Here is the crib for the example of fitting two phases.

1. Download the files
2. Plot the three $I(Q)$ files to compare them and to verify the phase content of the mixed data
3. Plot the three $G(r)$ files to compare them and to verify the phase content of the mixed data
4. Use PDFGUI to model the measured PDFs of silicon, nickel, and silicon-nickel mix
   (a) Within PDFGUI initiate three new fits, one each for silicon, nickel, and the silicon-nickel mix
   (b) Save the project (and make frequent saves as you work)
   (c) Generate the phases for silicon, nickel, and the silicon-nickel mix in the respective fits
   (d) Load the si.gr data file, the ni.gr data file, and the sini.gr data file in the respective fits
   (e) Set up the refinement conditions, and set the appropriate parameters to be refined as variables; pay special attention to the phase scale factors in the two-phase fit

  (f) Execute the refinements

  (g) Using the `Plot` utility of PDFGUI plot and inspect the fits

  (h) Save the project

5. Evaluate the refinement results to answer the scientific questions

5.6   Wait, what? How do I do that?

5.6.1   Download the files

Follow the instructions in Section 3.6.1.

5.6.2   Plot the three $I(Q)$ files to compare them and to verify the phase content of the mixed data

Start by making a plot of the three $I(Q)$ files on top of each other. Plot the files using your preferred plotting program. See Section 4.4 for help with .chi, .iq, etc., file formats. See Section 4.6.2 for help with doing this using Diffpy tools. Intensities across the three datasets have been coarsely normalized for easier comparison. The $I(Q)$ data have been intentionally truncated at 24 Å^{-1}.

  The data for silicon and nickel serve as a reference. By visually comparing them to the dataset of the silicon-nickel mixture, convince yourself that the mixture $I(Q)$ indeed contains reflections from the two ingredient phases and that no reflections remain unaccounted for.

5.6.3   Plot the three $G(r)$ files to compare them and to verify the phase content of the mixed data

Plot the three $G(r)$ files on top of each other using your preferred plotting program. See Section 3.4 for help with `.gr` file formats. See Sections 3.6.2 and 4.6.2 for help with doing this using Diffpy tools. The data for silicon and nickel again serve as a reference. By visually comparing them to the dataset of the silicon-nickel mixture, convince yourself that the mixture PDF indeed contains contributions from the two ingredient phases. Bear in mind that the apparent r-space resolution of the PDF data is not only affected by Q_{max} but by other factors as well. Think about what these factors are.

5.6.4   Use PDFGUI to model the PDF data of silicon, nickel, and silicon-nickel mix

Follow the steps described in Chapter 3 to set up three fits in the PDFGUI Fit Tree – one for silicon, one for nickel, and one for the silicon-nickel mix.

  For the pure Ni and Si samples, you can use exactly the same method we used in Chapter 3 for setting up the refinements and adding variables to the fit. First, you can use the `.cifs` to import the structures – after doing so, make sure that the program has correctly read the space group so that there are the right number of atoms in the structures. As we saw in Chapter 3, Ni forms a face centred cubic structure described by $Fm\bar{3}m$ with Ni at $4a$ (0,0,0) Wyckoff position with a site multiplicity of four. Silicon crystallizes in the diamond structure in space group $Fd\bar{3}m$ with Si at $8a$ $(0,0,0)$ that has a multiplicity of eight.

Next, import the data to the appropriate fit and specify the experiment and refinement conditions. When adding the experimental PDFs to the fits, remember to give the correct values for the instrumental parameters, here Q_{damp} as is provided in Table 5.4. You can also set the refinement range. As always, start with a narrow range, for example, 0–20 Å. Also make sure that the correct type of radiation (x-rays) has been specified and that PDFGUI has read the correct Q_{max} value from the $G(r)$ file.

It is now time to assign fit parameters and constraints, and specify the fit conditions for each of the two fits of the pure samples. For both Ni and Si, there are only two structural parameters: the cubic lattice parameter and the isotropic ADP of the atoms. In both of the structures, all atoms sit on special positions; hence the fractional coordinates are not to be refined. In addition, either δ_1 or δ_2 (but not both) should be assigned to take care of correlated motion effects, and we will also refine the overall phase scale factor. The total of four variables are assigned in the usual way using "@", within the **Phase Constraints** pane of PDFGUI (to get there, select the phase you want to edit in the fit tree, and then select the **Constraints** tab). It is important that each independent variable has been assigned a unique parameter. It is a good idea to review the fit setup prior to refinement execution. Once everything is set and verified, one proceeds with the refinement, where you should now get a nice agreement between data and model. Don't forget to save your project frequently. You can follow the steps we outlined in Chapter 3 for both fits.

You should now be ready to set up the multiphase refinement. PDFGUI allows you to have more than one phase per fit, as well as to have more than one dataset per fit. This makes it possible to carry out multiphase co-refinements on multiple datasets.

For the two phase refinement, the procedure is essentially the same as described above, except that there are now two phases to be added, the structure in space-group $Fm\bar{3}m$ and Si in $Fd\bar{3}m$, and the `sini.gr` measured PDF from the mixture. Just as before, you should define the parameters you want to refine as variables. This includes the lattice parameter, isotropic ADP, and δ_1 for Si and the lattice parameter, isotropic ADP, and δ_1 for Ni. We also need to set up the scale factors, as these ultimately determine the phase fractions. In the case of two phases, one should use a *maximum* of two independent scale factors. A simple way of doing this is to refine the two phase scale factors as independent parameters but keeping the overall scale factor (which can be set in the data constraints pane) fixed. Another method is to refine the overall dataset scale factor and constrain the two phase scale factors so that their values add up to one. Both scenarios lead to identical solutions, though the second approach gives more intuitive values for the mixture parameters, at least for a two-phase mixture. That gives a total of eight independent parameters, four for each phase. If you want to see how we set it up, as always, you can crib from our PDFGUI `.ddp` file in the **solutions** folder within the downloaded data. As the fit quality improves, increase the r-range of the fit to get more precise refined parameters.

Upon verification that everything is set up correctly, proceed with the refinement. Once the refinement is done, it is useful to visually evaluate the fit quality by utilizing the fit plotting utilities within the PDFGUI framework, as well as to save the project. As with any other application, saving the project frequently as you go is a wise thing to do.

5.6.5 Evaluate the refinement results within the PDFGUI suite to answer the questions

Once the refinements are completed, review the results by selecting the desired fit and choosing the `Results` pane, in which the fit summary is provided. In the case of a single phase refinement, values of both refined and fixed parameters will be summarized, followed by the quantitative measures of the fit quality. The `Results` pane can be scrolled up and down. In the case of a multi-phase refinement, the summary will contain information on all ingredient phases in the model, as well as the relative phase fractions displayed in terms of phase fractions per atom, unit cell, and mass. Bear in mind that the phase fraction is not simply a ratio of the respective scale factors, as the PDF in general depends on both the scattering properties of the atoms in the ingredient phases and the relative number of atoms in the unit cells of the ingredient phases. But the fractions in terms of physical quantities, such as mass fraction, mole fraction, etc., are computed by the program and can be found in the `results` pane of the fit (select the fit by left clicking the fit name to select it and then selecting the results tab).

5.7 Problems

1. Why are some structural parameters for the mixture sample quite comparable to their single phase references, and some show discrepancies? What factors affect the discrepancies?
2. What was the multiphase sample in this chapter comprised of? Could you determine the phase fractions in this sample without PDF and why?
3. What is the key difference between the Rietveld and PDF multiphase approaches? What are advantages and disadvantages of each of the methods?
4. Describe where the two approaches are applicable, and comment on the reliability of these approaches.
5. Think of some scientific cases where multiphase PDF modelling provides unique insights.

5.8 Solution

Table 5.3 summarizes the results we obtained and also provides a comparison of the refined structural parameters of the mixed phases with those of the single phase references.

By now you should have the necessary information to answer the questions from the beginning of the chapter. The mixed sample was prepared by carefully mixing silicon and nickel in a 9 to 1 mass ratio, as indicated in the table. By refining the

Table 5.3 Results of a two-phase PDF refinement from x-ray data of an admixture of Si and Ni powders, and of single-phase refinements on the pure materials. Refinements were carried out using the PDFGUI suite from 1.5 Å to 20.0 Å, with Q_{damp} set to 0.05 Å^{-1}. The Nominal column shows the expected (nominal) mixture ratios from the known amounts used in the mixture. The Mixture and Pure columns correspond to the results of the refinements on the PDFs from the admixture and the pure compounds, respectively.

Phase	Si			Ni		
Refinement type	Nominal	Mixture	Pure	Nominal	Mixture	Pure
Mass fraction	0.9	0.887(3)	1.0	0.1	0.113(3)	1.0
Atomic/mole fraction	0.95	0.942(2)	1.0	0.05	0.058(2)	1.0
Unit cell fraction	0.9	0.891(3)	1.0	0.1	0.109(3)	1.0
Lattice parameter (Å)	-	5.4294(7)	5.4296(5)	-	3.5217(6)	3.5239(2)
U_{iso} (Å^2)	-	0.0095(3)	0.0094(2)	-	0.0055(3)	0.00688(7)

models over ≈ 20 Å range, the program returned a mass fraction that is close to the target phase ratio. We also include in the table the atomic fraction (one of the quantities returned in the results section of the fit). Because the constituents in the sample are simple elements, this also corresponds to the mole fraction in this case. We can see that the sample is actually 95% silicon by mole, and yet the refinement was clearly able to detect Ni and return an accurate phase fraction. The scattering power in the signal scales with the square of the atomic number for elements, and since the ratio of the atomic numbers is $28/14 = 2$ in this case, the actual nickel signal will be four times stronger per atom, which helps us to see this small amount of Ni.

The program also returns the unit cell fraction of the phases. This is the ratio of unit cells of each phase. In this case around 10% of the unit cells are nickel. Since there are twice as many silicon atoms in a silicon unit cell (8) than nickel atoms in a nickel unit cell (4), this means that only 5% of the atoms are nickel, as we found from the atomic fraction. This unit cell fraction measure is particularly useful if you are interested in molecular species and you want to compare the relative number of molecules. Regardless of the size of the molecules, their composition, and their molar mass, you can compare the numbers of molecules in each phase by using the unit cell fraction and correcting for any difference in the number of molecules in each cell. If each cell contains one of its respective molecules, the unit cell fraction is the actual ratio of molecules. If (as here, where the "molecule" is a single atom) one phase has twice as many molecules as the other phase, you simply multiply the unit cell fraction measure by the ratio of the number of molecules. This also works if we are interested in the relative number of formula units in an inorganic material.

In Table 5.3 we see that the unit cell fractions and the weight fractions are almost identical (and, within the precision given by the number of significant digits, actually are identical for the nominal mixture). This is an accident of nature for this particular case and won't be true in general. Here the ratio of the number of formula units is 2, and the mass ratio is also almost exactly 2.

As for the comparison of the refined structure parameters from the phases in the mixture compared to those from the reference single phase samples, Table 5.3 provides the side-by-side comparison. What did you get? It is important to note that silicon is the majority phase in the sample. Despite Ni being a two times stronger scatterer than Si, the mixture is dominated by the Si signal. The refined Si parameters in the mix are rather robust and comparable to the single-phase reference. The parameters pertaining to the minority nickel phase are also in reasonable agreement with the reference but are known with less certainty and are more susceptible to any experimental artefacts and data imperfections that propagate to the PDF. Therefore the Ni parameters obtained from the mix less reliably reproduce the values obtained for the single phase Ni reference. The discrepancy is relatively small though. Notably, the example used here represents a simple case of pedagogical value. In real materials, multiphase situations are often much more complex, with the ingredient phases being of much lower symmetry. Also, due to strains and other such effects, the parameters of the ingredient phases do not necessarily have to match those of their single-phase references, if such exist. However, useful insight can still be gained by making such comparisons.

If considering the real-space PDF analysis done here versus a "standard" Rietveld analysis in reciprocal-space, the Rietveld approach treats the average crystal structure in reciprocal space, whereas PDF is sensitive to local structure. The Rietveld multiphase approach therefore handles mixtures of independent bulk phases quite well, even when minority bulk phases are present only in small single digit percentages, providing there is enough counting statistics for the weak signals to rise above the background intensity. The sample we deal with in this chapter comprises a mixture of micron-sized grains of Si and Ni. As such, we are dealing with a mixture of two *bulk* phases that are completely independent and that both span the micron length scale. Therefore, there is nothing unique in applying PDF analysis to this problem, and a standard reciprocal-space Rietveld analysis would work in this case. In fact, when dealing with bulk crystalline phases with well-defined long range order, data treatment in reciprocal-space is often more succesful, as unique reflections of different participating phases have a better chance of being observed and differentiated than in the PDF. In real-space, depending on the complexity of the structures in play, the PDF peaks could become overlapped fairly quickly as r increases. However, the Rietveld approach cannot adequately handle phase mixtures where one or more phases is nanoscale or amorphous in character. In fact, the presence of such a phase, whose signal may be completely diffuse, is often tossed away in Rietveld refinement together with the experimental background, and here PDF can be extremely useful for determining the presence and quantity of phases that are not crystalline.

In both reciprocal- and real-space methods, the uncertainties of the results are associated with the scattering contrast of the ingredient phases and the relative abundance of the phases, as well as any refinement instabilities, such as correlations between the refined parameters. In the case of reciprocal-space Rietveld refinement, the scale factors that affect quantification can be correlated with the parameters used to describe the background. In PDF this is of lesser concern because the background signal is explicitly and independently measured and subtracted from the data. However, the reliability of the phase fractions determined from PDF could be affected by correlations between the scale factors and the atomic displacement parameters, particularly when dealing with complex low-symmetry structures and when refining over narrow r-ranges. A parameter correlation matrix is reported in PDFGUI in the `Results` pane of the `Fit Tree` (select the fit you are interested in then click on the results tab). One should always inspect this once the refinement has been completed. If variables are reported with very high (> 0.9) correlations, one could consider reducing the number of parameters, for example going from anisotropic ADPs to isotropic ADPs or extending the r-range of the refinement. We recommend that you explore this aspect by refining the two-phase model to silicon-nickel mixture data over various r-ranges from very short to extended, and observe how this affects the correlation of the refined parameters.

PDF multiphase modelling provides an opportunity to study the structure of many different systems. For example, core-shell nanoparticle complexes may be approximated by treating the core and shell as independent structural phases, assuming that these are sufficiently distinct (Lappas *et al.* 2019; Antonaropoulos *et al.* 2022). This approach gives useful insights in cases of nanoparticles with significantly thick cores and shells. In such a case the problem reduces, on at least some nanometer/subnanometer lengthscale, to a problem of handling two independent bulk-like phases. If the particles get to the ultra-small size regime (with coherence lengths of less than 3 nm), the data cannot always be treated with this approach – more complex structures may arise due to cross-correlation of the core and shell parts of the system, the existence of an interface layer between the two, surface relaxation effects, and a significant amount of strain, which may render crystallographic structural models inadequate. However, useful insights can still be gained from our traditional "real-space Rietveld" small box approach in these cases. When needed, such complexities can be handled by modelling these experimental PDFs using DIFFPY-CMI or a big-box modelling method where the core-shell nanoparticle system may be treated explicitly. We will introduce this kind of data analysis in Chapter 11 of this book.

5.8.1   On reported uncertainties from PDF refinements

In Table 5.3 the estimated standard deviations (e.s.d.'s) returned by the program are shown in parentheses after the values of the refined parameters. In PDFGUI you can find all the values and their e.s.d.'s by looking in the results tab (select the fit title and then the results tab). The values of the refined variables along with the estimated errors on those numbers are reported to a fixed number of significant digits in there. It is common practice to round the standard errors to one significant digit as we have done in Table 5.3. In the table we use the standard practice of putting this rounded error value in parentheses and rounding the refined parameter value to the number of significant digits such that the number in the parentheses is next to the uncertain number in the parameter value. That is, this number (just to the left of the parentheses) will be higher or lower than what is reported by less than the amount in the parentheses around two thirds of the time and within twice this value 96% of the time, if we do the identical measurement multiple times. The rounding is maybe best illustrated with an example. If a variable $V = 0.068973$ and its standard error is reported as 0.000379, we would report this as 0.0690(4). There is one exception, when the first two significant figures of the e.s.d.'s are between 10 and 14, where it is common to keep the e.s.d. to two significant figures. Again as an example, if we had $V = 0.068973$ with an e.s.d. of 0.000139, we would report this as 0.06897(14).

However, the uncertainties reported in Table 5.3 deserve some discussion, which we do below, as there are complications that are not widely appreciated. As trainee scientists we are told that we should always assess the uncertainty in our measurements and report these uncertainties. Furthermore, we are taught how to do error propagation so that if we know the uncertainties on some quantity, we know how to propagate them to a derived quantity. There are standard approaches for doing this in least-squares regression problems such as PDFGUI is using. The error propagation has therefore been incorporated into the PDFGUI code and will take whatever input uncertainties it is given and propagate them through the regression to estimate uncertainties on the refined parameters. It is beyond the scope of this book to discuss this in detail. However, danger lurks in the PDFGUI results pane, which automatically reports numbers for the estimated uncertainties regardless of whether meaningful uncertainties (or any uncertainties at all) were handed to it with the measured data, for example, in the .gr file. Modern data acquisition methods, such as the *RA-PDF* method, are highly complex. For example, they may involve converting x-ray absorption events in a phosphor into cascades of optical photons that may then go down an optical fiber and impinge on a CCD detector, or a photomultiplier device, and so on and so forth. Almost all the assumptions of standard error estimation and propagation are violated in these processes, making it practically impossible to use standard methods to get meaningful uncertainty estimates on measured intensities, even in the 1D diffraction patterns themselves. This is not particularly a PDF issue, because we know how to propagate errors through the Fourier transform (Toby and Billinge, 2004), but cuts across most of the world of diffraction. It is the subject of current research and probably deserves a book all to itself, so we won't discuss it too much here. But it means that in nearly all cases, the uncertainties reported by PDFGUI are at best problematic and at worst completely misleading. For this reason, in our work we often don't report them at all.

In the current case, we have reported them for illustration purposes and to invoke this discussion. Actually, if you open the Ni.gr file you will see that *no uncertainties* are reported in there, yet PDFGUI has managed to "estimate" an uncertainty on the refined parameters. How is this even possible? What PDFGUI does is to replace all the uncertainties it can't find with the value "1" and then propagates those. This seems like it would result in incorrect refinements, let alone incorrect error estimates, because we usually weight points in the refinement based on their estimated uncertainties. In the case of the PDF, the uncertainties are almost uniformly distributed among the points (Toby and Billinge, 2004), so giving equal weighting to each point will result in a correct refinement – i.e. the values of the refined parameters will be valid, but estimates of the uncertainties are uncertain. More work is needed to properly understand all the issues with estimating uncertainties in diffraction refinements, but we note one final point that it is common with data from synchrotron measurements from large area detectors: the fit residuals are actually dominated by model errors (for example, slightly incorrect line-shapes) and not by the statistics. Two concepts are sometimes distinguished in the literature by differentiating parameter "precision" (i.e. how precisely did we estimate the parameter given that we made a measurement with an imperfect statisitical sampling) from model "accuracy" (i.e. how close to the actual value is our estimate). The estimated errors that are reported in nearly all tables in diffraction and PDF papers report an estimate of precision (when they are coming from properly estimated and propagated statistics, which they often aren't). They do not report the accuracy. They are only estimates of the real accuracy on the measured parameter values if the model is completely correct, which is rarely the case. Our studies of this issue suggest that most current synchrotron diffraction measurements are in a regime where the parameter value uncertainties are dominated by model errors and not by the statistics, which further makes the reporting in tables of standard errors propagated from the statistics problematic. Treat these numbers with a great deal of care, especially if there is not a significant discussion in the paper of how they were arrived at.

As a matter observation, the errors reported by PDFGUI seem to be giving "reasonable" error estimates, in the sense that they report significance that is at around the right order of magnitude (the right number of significant digits) based on our intuition. This may be a happy accident though, as it is doing it based on mathematics for propagating statistical errors without any knowledge of the statistical errors themselves. However, they should be treated with caution. They do contain useful information in the sense that they do a reasonable job of estimating the *relative* uncertainties between the different refined variables. This is a measure of the curvature of χ^2 as a function of a variation of the parameter value, which depends on the physical model and its relation to the information content in the signal and does not depend on the statistics; hence it is reliable even in the absence of the correct statistical information. So from Table 5.3 we see that the PDFGUI-reported errors suggest that we know the lattice parameters to four significant digits, but the phase fractions to three and the ASDs to no better than two. These relative weightings are probably valid, but the actual numbers themselves should not be over-interpreted.

5.9   DIFFPY-CMI solution

This section describes the DIFFPY-CMI solution. It may be skipped on a first reading, but it will be helpful to you when, in the future, you want to learn how to use DIFFPY-CMI. To get started with DIFFPY-CMI, follow the instructions in Section 3.9. The DIFFPY-CMI code for the solution will be in a Python file in the diffpy-cmi folder under solutions for this chapter in your downloaded data. Read through the comments there. They are expanded on below in greater detail where needed.

5.9.1   `fit2P.py`

This file presents a DIFFPY-CMI solution to the refinement of the PDF collected from the two-phase sample that you did using PDFGUI above. To run this code, you should have downloaded the script in a directory called DIFFPY-CMI, with a parent directory called **solutions** and another directory called **data** located at the same directory level as **solutions**. The **data** directory should contain the files **sini.gr**, **Ni_Fm-3m.cif**, and **Si_Fd-3m.cif**. Have a look at **fit2P.py**. Lines beginning with **#** are comments.

#1-7:   As we now have two phases, we need to provide initial values for the parameters of both phases, and we need to give the location of two structure files.

 # 8-13:   In the `make_recipe` function (see Section 3.9), we now need to provide the name and location of a PDF data file and *two* structure files, one for each structure in our two-phase fit. We load in each structure file as well as the experimental PDF data. We create two unique profile generators, each with its own unique name and structure, G_Si and G_Ni for Si and Ni, respectively. We also create a fit contribution and give it our profile instance, which contains our experimental data as well as both our profile generators.

14:   In this example, we know that our sample contains two phases, and we'd like to compute a model PDF signal that has a contribution from each phase. Here we can see the power and flexibility of using multiple profile generators managed by a single fit contribution. We define our contribution equation as s2 * (s1_Si * G_Si + (1.0 -s1_Si) * G_Ni). Here, s2 is a global scale parameter, similar to the data scale in PDFGUI. G_Si and G_Ni are names we have given to the PDF generators for Si and Ni, respectively. s1_Si represents a phase fraction scale parameter for Si, and since we know that we want the fraction of Ni and Si to sum to one, we multiply the Ni PDF G_Ni by (1.0 - s1_Si).

15-16:   We create a fit recipe, assign our fit contribution, add the two scale parameters s1_Si and s2, and initialize these parameters with values defined in previous steps.

17:   In this part of the script we add fit constraints. This is a new step that you haven't seen before. Fit constraints allow us to apply limits or boundaries to parameter values by modifying the objective function we will minimize. They can help guide a fit away from non-physical parameter values.

In this case, we know that both s1_Si and s2 should always be positive, because a negative PDF signal is not particularly helpful here. We also know that s1_Si should always be less than 1.0, so as to keep (1.0 - s1_Si) positive. The `recipe.restrain` function helps us with this: we pass it the name of the parameter (or any string expression involving defined parameters) we want to constrain and some keyword arguments. Specifically, `lb` and `ub` represent the lower and upper bound we'd like to impose. Keep in mind that we don't have to impose both! `sig` describes the shape of the boundary, and lower values represent a sharper boundary, while larger values represent a softer boundary. We also switch the constraint to be scaled, using the `scaled` keyword argument. This ensures that the constraint contribution to the total fit residual is roughly as significant as any other data point throughout the fit.

18-20:   Just as we've seen before, here we add space group symmetry constraints and a correlated motion parameter (δ_1). Keep in mind that we need to do this for both PDF generators, one for each phase. We make use of some Pythonic looping to do this without duplicating too much code. After this, everything is ready, and we return our recipe from the function.

21-26:   The remainder of our code proceeds identically as in the previous examples. We just need to be sure to pass our **make_recipe** function the full path to both structure files we will use.

6

More advanced crystal structure modelling: The room-temperature structure of crystalline $Ba_{0.7}K_{0.3}(Zn_{0.85}Mn_{0.15})_2As_2$

Benjamin A. Frandsen

6.1 Introduction and overview

The goal of this chapter is to carry out a structural refinement on a crystalline material that is more complex, with more degrees of freedom, than the nickel and platinum that we looked at in Chapter 3 and the mixed phase from Chapter 5. We will build and refine structural models for $Ba_{0.7}K_{0.3}(Zn_{0.85}Mn_{0.15})_2As_2$, a complex crystalline alloy. The analysis includes how to model an alloy in the virtual-crystal approximation (different atoms on the same crystallographic sites) and how to search for local structural distortions.

6.2 The question

$Ba_{0.7}K_{0.3}(Zn_{0.85}Mn_{0.15})_2As_2$ has been the subject of recent study because it is a dilute ferromagnetic semiconductor (DFS) with interesting magnetic and transport properties (Zhao *et al.*, 2013; Frandsen *et al.*, 2016b) that have potential applications in spintronics (Dietl and Ohno, 2014). Do not worry if you are not familiar with DFS materials; you can still get through the chapter and do the structural analysis without any problem. Preliminary characterization (Zhao *et al.*, 2013) showed that the average crystal structure of $Ba_{0.7}K_{0.3}(Zn_{0.85}Mn_{0.15})_2As_2$ (shown in Figure 6.1) has tetragonal symmetry and that the K and Mn atoms are distributed randomly across the Ba and Zn sites, respectively. We would like to understand more detailed aspects of the structure, such as the average composition (i.e. are the Ba/K and

Atomic Pair Distribution Function Analysis. Benjamin A. Frandsen, Oxford University Press.
© Simon J. L. Billinge and Kirsten M. Ø. Jensen (2023). DOI: 10.1093/oso/9780198885801.003.0006

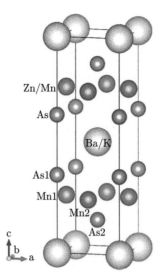

Figure 6.1 Average crystallographic structure of $Ba_{0.7}K_{0.3}(Zn_{0.85}Mn_{0.15})_2As_2$ determined from x-ray diffraction.

Zn/Mn concentrations really what we think they are?), the precise positions of the atoms within the unit cell, and whether any evidence exists for relaxational disorder, where atoms displace off their average positions, or other forms of local symmetry breaking.

We have several scientific questions related to the structure of $Ba_{0.7}K_{0.3}(Zn_{0.85}Mn_{0.15})_2As_2$, including how the structure changes as it is cooled into the ferromagnetically ordered state at low temperature, whether there are any differences between the local and average structure, and how these differences relate to the material's magnetic and transport properties. However, in this chapter, we will focus just on refining the structure from the PDF data at a single temperature.

6.3   The result

From our PDF analysis, we confirmed that the atomic structure of the sample on a length scale of 30 Å is in excellent agreement with the tetragonal structure suggested from previous diffraction measurements. Knowing the structure in detail provides a crucial starting point for understanding the magnetic and electronic properties. As a pleasant surprise, we also found evidence for a symmetry-broken local structure in the form of short-range orthorhombicity and positional displacements (keep your eye out for this as you work through the example!). These results, as well as a more detailed analysis of the temperature dependence of the average and local structure, are discussed more fully in Frandsen *et al.* (2016b).

6.4   The experiment

The experiment was performed on a finely ground powder sample of $Ba_{0.7}K_{0.3}(Zn_{0.85}Mn_{0.15})_2As_2$ (nominal composition) under the conditions given in Table 6.1. The files to download are listed in Table 6.2.

Table 6.1 Experimental details.

Facility	NSLS at Brookhaven National Laboratory
Beamline	X17-A
Date collected	September 2014
Detector type	Perkin Elmer amorphous silicon 2D detector
Sample geometry	Powder in 1 mm ID Kapton capillary
Sample environment	Room temperature, ambient conditions
Total exposure time	120 s
Wavelength	0.18597 Å
Q_{max}	25.0 Å^{-1}
Q_{damp}, Q_{broad} from calibration	0.03842 Å^{-1}, 0.01707 Å^{-1}

Table 6.2 Files to download.

Filename	Note
`BaZn2As2_K-Mn-doped_300K.gr`	Experimental PDF data
`BaZn2As2.cif`	CIF file for parent compound $BaZn_2As_2$

6.5   What next?

1. Download the files.
2. Get a CIF file from PDFITC to use as a starting point.
3. Open up the file `BaZn2As2.cif` in a text editor, and modify the section with the atomic positions and occupancies to reflect the nominal sample composition.
4. Create a PDFGUI project and start a new fit.
5. Create the structure for the fit by loading the edited `.cif` file, load the PDF file, and set the fit range to a rather standard $1.5 \leq r \leq 30.0$ Å.
6. Plot the PDF to ensure that it looks reasonable.
7. Fix the instrumental parameters to appropriate values, and assign the refinable parameters to variables, including the overall scale factor, lattice parameters, atomic positions, ADPs, atomic occupancies, and δ_1 or δ_2 . You

can use the space group information to load the appropriate atomic positions and ADP variables automatically.

8. Perform a calculation with the starting structure to compare with the data and make sure it is at a good enough starting point. You may use this to estimate a sensible starting scale factor.

9. Refine your model to obtain the best fit. Begin by only refining the parameters with the largest influence on the PDF (e.g. scale and lattice parameters) and then slowly free up more variables in the model.

10. Inspect the fit (you should be doing this along the way), and verify that it is of good quality, similar to Figure 6.2.

6.6 Wait, what? How do I do that?

6.6.1 Download the files

Follow the instructions in Section 3.6.1.

6.6.2 Get a CIF file from PDFITC to use as a starting point

Using STRUCTUREMINING (Yang *et al.*, 2020) at PDFITC (Yang *et al.*, 2021) is a great way to rapidly find one or more relevant CIF files to start your modelling campaign. Take your experimental PDF (`BaZn2As_K-Mn-doped_300K.gr`), and upload it to the STRUCTUREMINING app at PDFITC. You have to log in to use the app. Click on the STRUCTUREMINING tile, and upload the $G(r)$ file. You can give a list of elements that you want the app to search over. For example, Ba-Zn-As-* will search the database for compounds that contain all of Ba, Zn, As, and anything else (or nothing else). STRUCTUREMINING will compare the uploaded PDF to all the filtered PDFs that it finds from the database and return a rank-ordered list of candidate structures. You can plot these to see the agreement and download the CIF files of any that you want to investigate further. Save the CIF file into your working directory. For your convenience, we also provide a relevant CIF file in the **data** directory, but it is recommended to practice the STRUCTUREMINING workflow as you will find it helpful in the future.

6.6.3 Modify the CIF file

As already discussed briefly in Chapter 3, Crystallographic Information Files (CIF files) are a standard way to report crystal structures – they are used by almost all crystallographic software, including PDFGUI. Here, we will also use a CIF to set up our structure, but instead of taking it directly from a database and using it in PDFGUI, we first need to modify it to reflect the nominal composition of our sample. The CIF you have downloaded may have the right composition $(Ba_{0.7}K_{0.3}(Zn_{0.85}Mn_{0.15})_2As_2)$, but more likely it will not. For example, the CIF file in the downloaded data reflects the structure of $BaZn_2As$, which is isostructural to our sample but has a different composition. Thus, we need to introduce some K on the Ba sites and some Mn on the Zn sites in the crystal structure.

This may be done from within PDFGUI. However, an alternative approach is to create a modified CIF file with the correct composition and load this directly into PDFGUI. Since this is a useful trick, here we describe how to take this approach, using the CIF file that was supplied with the downloaded data. If you downloaded a different file from PDFITC, the approach will be the same, but details of the edits you need to make will depend on precisely the CIF file you chose to download. In every CIF, you will find lines that indicate the type of atom, its position in the unit cell, the symmetry characteristics of that position, its occupancy, and sometimes additional information such as the ADP. In the $BaZn_2As_2$ CIF we supplied, you will find this information in the "loop" block spanning lines 84–98. The lines beginning with an underscore character specify the meaning of the entries in the lines that do not start with an underscore character, i.e. "Ba1 Ba0+ 2" and so forth. We need to modify this block to include K on the Ba site and Mn on the Zn site, with appropriate occupancies reflecting the nominal composition. To do this, copy the entire line starting with "Ba1" and paste it in a new line directly below. Change the atom label and type to K, but do not change any of the symmetry or positional information. Change the occupancy (second-to-last entry) for Ba to 0.7 and for K to 0.3. Now, do the same thing for the Zn line: copy and paste it directly below, update Zn to read Mn, and change the occupancies to 0.85 for Zn and 0.15 for Mn. Additionally, to avoid any potential problems when reading in the CIF file to PDFGUI, delete the "0+" characters from the second column of each row, since the structure reader in PDFGUI sometimes has problems with these. Finally, you may also add in the K and Mn atoms to the ADP loop at the very bottom of the CIF, although this is not crucial. If you do this, then give the K atoms the same ADP values as Ba and the Mn atoms the same values as Zn, since these atoms are sharing sites. You may now save the modified CIF.

6.6.4   Create a PDFGUI project and start a new fit

Open up PDFGUI and create a new fit. Refer to the earlier description in Section 3.6.3 if you need a refresher on how to do this in PDFGUI. Give your fit a descriptive name and save the project for safekeeping. Don't forget to make frequent saves of your project for a quicker recovery from crashes.

6.6.5   Load the structure and data files into your fit

Referring to Section 3.6.4 if necessary, load the structure from your modified CIF. Verify that the phase information has been populated correctly; you can make any necessary changes directly within PDFGUI. Add the dataset in the same way as outlined in Section 3.6.3, and verify that it has read the correct scatterer type (x-rays) and Q_{max} value in the **Configure** panel. Change the fit range to be from 1.5 to 30 Å.

6.6.6 Fix the instrumental parameters to appropriate values and assign the refinable parameters to variables

In the dataset configuration tab, set the appropriate values for Q_{damp} and Q_{broad} (listed in Table 6.1).

Now we will set all of the parameters that will be refined during our modelling of the PDF as variables. First is the scale factor, which you can set either in the dataset configuration tab or the structure configuration tab. We recommend to do it in the data configuration tab. We also want to add the structural parameters as variables. Compared to the simple Ni, Pt, and Si structures we have dealt with so far, the $Ba_{0.7}K_{0.3}(Zn_{0.85}Mn_{0.15})_2As_2$ structure is more complex, and we will therefore have more structure variables. Left-click on the phase in your fit tree to select it, and click on the `Constraints` tab. Set the lattice parameters as variables, making sure to use a different number after the @ for all independent variables. Remember to take into account that the cell is tetragonal when you define the variables. To set the parameters for the atomic positions and ADPs, we can take advantage of a shortcut using the space group information. When in the `Configuration` tab, click on the `elem` column to highlight all the atoms, then right click on the highlighted region to get the Constrain by space group dialogue, and select `Symmetry constraints`. From the menu that appears, make sure the space group is correct ($I4/mmm$ in this case), and select the two options for constraining the positions and temperature factors. Click `OK`, and it will automatically define the appropriate variables that respect the space group symmetry.

Finally, we need to set the atomic occupancies as variables. Still in the phase pane, in the column labeled `occ`, select the two boxes corresponding to the Ba atoms, and set them as a variable. Because the Ba and K atoms are distributed across the same crystallographic site, their occupancies must add up to unity for full occupancy of the crystallographic site. Therefore, the appropriate constraint for the K atoms is 1 minus the Ba occupancy. To enforce this, highlight the two `occ` boxes corresponding to the K atoms, and type "1-@#", where you use the same parameter number (#) as for Ba. Repeat this procedure (using a different parameter number) for the Zn and Mn atoms. If you want to see how we set up the refinement, you can peek at our `.ddp` solution file distributed along with the data, but we highly recommend trying it all by yourself!

6.6.7 Perform a starting calculation

Just as we did for Ni in Section 3.6.7, we will begin by calculating what the PDF from the starting structure will look like. Select your fit, go to the `Calculations` tab at the top of the window, and select `New calculation`.

The calculation will appear in the fit tree. Select the calculation, and press the blue gear icon to do the calculation. To view the calculation, select the calculation from the fit tree and then press the `Plot` icon on the top task bar. You can also plot the data by selecting the data from the fit tree and pressing the same plot icon. Visually compare the calculation and data to make sure they look approximately the same. You can also compare the magnitudes to get a reasonable starting value for the scale factor, which may help the initial fit converge better. In this case, we notice that the magnitude of the calculated PDF is about three times larger than the data, so we can set the initial scale factor to 0.33.

6.6.8. Refine the model

You are now ready to refine the parameters of the model. If necessary, refer back to Section 3.6.7 for a reminder about how to run refinements in PDFGUI. The refinement process works best when you take a sequential approach: first refine the parameters that have the biggest effect on the calculated PDF, then add more and more variables until you have the best possible fit. You can fix and unfix variables using the checkboxes in the fit pane, and you can copy the refined variable values to the initial parameter values for each new cycle of refinement in the way described in Section 3.6.7. If you need a hint for the refinement sequence, see the Solution section.

6.7   Problems

1. What happens if you try to refine all of the parameters together on your first refinement cycle?
2. How does the quality of the fit at low r compare to that at high r? What is the significance of this in terms of the local structure of this material?
3. Given that this structure is tetragonal, what should the relationship be between the a and b lattice parameters? What values must α, β, and γ take? What happens if you let all of the lattice parameters vary independently (thus allowing the tetragonal symmetry to be broken) during your refinements?
4. Fix the Ba/K atomic occupancies to values far away from the nominal composition, and observe the effect on R_w. Do the same for the Zn/Mn occupancies. Which has a greater effect on the fit quality? Why? What does this tell you about the ability of x-rays to reliably distinguish between elements that are close to each other in the periodic table?

6.8   Solution

In Figure 6.2, we display the best fit for $Ba_{0.7}K_{0.3}(Zn_{0.85}Mn_{0.15})_2As_2$, which we obtained after refining the scale factor, lattice parameters, general atomic positions, ADPs, δ_1, and atomic occupancies.

To arrive at this fit, we performed the refinement in a sequential way. In the first iteration, we refined just the overall scale factor and the lattice parameters (with a and b set to the same parameter to respect the tetragonal symmetry, and with all the angles left fixed at 90°), leaving all other parameters fixed. These parameters have the greatest influence on the calculated PDF, so it is important to get them to approximately correct values before refining additional parameters. Visually inspecting the fit at this point verified that we had the right basic structure,

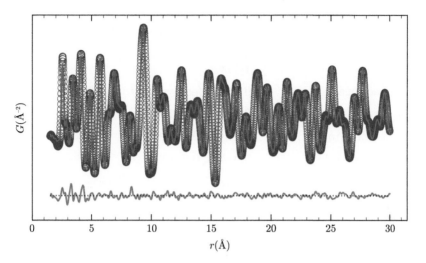

Figure 6.2 Optimal PDF fit for $Ba_{0.7}K_{0.3}(Zn_{0.85}Mn_{0.15})_2As_2$ obtained by refining tetragonal lattice parameters, general atomic positions, anisotropic ADPs, and atomic occupancies over a range from 1.5 Å to 30 Å.

since the calculated and observed peaks in the PDF lined up reasonably well. After copying the refined parameter values to the initial values, we then freed up the parameters for the atomic positions. Because of the symmetry of the structure, there is actually just a single parameter to refine – the z-position of the As atoms. Next, we added the ADP parameters to the refinement, and after that, δ_1 (see Section 3.8 if you need a reminder about what this parameter represents). Note that we could also have used δ_2, and there would have been little to no difference in the overall quality of fit. In practice, just choose either δ_1 or δ_2 and stick with it, but do not refine both parameters, since they are very highly correlated. Finally, we refined the atomic occupancies. The refined parameters are provided in Table 6.3.

If we had not performed the fit in this sequential way, then there is a good chance the refinements would not have worked well or at all. If there are many parameters being refined simultaneously and a significant portion of them are well away from their optimal value (as is often the case when you are starting a fit), then the refinement algorithm can often choke or get stuck in a local minimum far away from the best fit. For that reason, it is important to refine the parameters in a sensible order, such as that presented here. Try different strategies by changing starting parameter values and the order of refinement to check for stability (do you get the same result from different starting values?) and to try and get out of local refinement minima.

If you turn on (unfix) a variable and it doesn't move *at all* (i.e. it doesn't change in any of its significant figures) in the refinement, the variable may be stuck. Try changing its value slightly and try again.

On inspecting the fit residual in Figure 6.2, you will notice that the fit quality is significantly worse at low r (below about 5–7 Å) than at high r.

Table 6.3 Refined parameters from the PDF modelling of $Ba_{0.7}K_{0.3}(Zn_{0.85}Mn_{0.15})_2As_2$.

a (Å)	4.13815
c (Å)	13.5306
z for As (fractional)	0.63711 (equivalent to 0.36289, 0.13711, 0.86289)
U_{11}, U_{33} for Ba/K (Å^2)	0.01297, 0.01945
U_{11}, U_{33} for Zn/Mn (Å^2)	0.01388, 0.01948
U_{11}, U_{33} for As (Å^2)	0.01133, 0.01108
Refined composition	$Ba_{0.6745}K_{0.3255}(Zn_{0.7728}Mn_{0.2272})_2As_2$
δ_1 (Å)	1.578
Scale factor	0.3906
R_w	0.0893

This is always exciting to a PDF person, because it means there may be some interesting local symmetry breaking on a short length scale – while the long-range, average structure is apparently very well described by the expected tetragonal structure, the local atomic correlations within the first several Ångstroms are evidently rather different. In fact, the local structure hinted at by the fit residual actually became the main focus of a separate paper, where it was shown that the lattice becomes locally orthorhombic and certain atoms relax off their average positions on a short length scale (Frandsen *et al.*, 2016b).

For the moment, though, we will ignore the failure of the structural model at very low r and instead focus on the success of the model at longer r.

The excellent quality of the fit reassures us that we have a very accurate model of the structure, but it is still advisable to go a bit further to test the reliability of the fit. First, we can look at the tetragonal symmetry of the lattice parameters. If we assign a new parameter to the b lattice parameter so that it is not constrained to be the same as a, we can do the refinement again and see if a and b converge in accordance with tetragonal symmetry. Indeed, they do converge to be within 0.001 Å even if we give them rather different starting values. Likewise, if we allow the unit cell angles to vary, we see no tendency for them to refine to values different from 90°. Therefore, on the 30 Å length scale of these fits, there is no evidence for a breaking of the average tetragonal symmetry of the lattice.

We can also examine the refined atomic occupancies. The refined values are fairly close to the nominal composition, which is reassuring for both us and the sample growers. To get an idea of how reliable the refined composition is, we can fix the occupancies to values far away from the optimal value and observe the effect on the overall fit quality. If we fix the Ba concentration to a much lower value, say 0.35, then we see that the fit becomes noticeably worse, with R_w increasing from approximately 0.089 to 0.122. On the other hand, if we fix the Zn concentration to

0.35, R_w only increases to about 0.093, so the Zn/Mn occupancy has a much smaller effect on the fit than the Ba/K occupancy (although still a noticeable enough effect to feel fairly confident in the refined values). This difference between Ba/K and Zn/Mn is exactly in line with what we should expect, since the x-ray scattering strength of an atom is directly related to the atomic number. Therefore, x-rays are most capable of distinguishing between elements that are rather widely separated on the periodic table, such as Ba and K, rather than elements much closer on the table, such as Zn and Mn. This is of course not the case for neutrons, for which the scattering strength is more or less randomly distributed across the periodic table. This is one reason that x-rays and neutrons are so complementary – they provide very different scattering contrasts.

Congratulations! At this point, you have learned most of the essential skills for doing PDF refinements of crystalline materials. We hope you put these skills to good use!

6.9 DIFFPY-CMI **Solution**

This section describes the DIFFPY-CMI solution. It may be skipped on a first reading, but it will be helpful to you when, in the future, you want to learn how to use DIFFPY-CMI. To get started with DIFFPY-CMI, follow the instructions in Section 3.9. The DIFFPY-CMI code for the solution will be in a Python file in the diffpy-cmi folder under solutions for this chapter in your downloaded data. Read through the comments there. They are expanded on below in greater detail where needed.

6.9.1 Running DIFFPY-CMI examples

See the instructions in Chapter 3 for help on running Python files.

6.9.2 `fitCrystalGen.py`

This file presents a DIFFPY-CMI solution to the refinement of the PDF collected from the $Ba_{0.7}K_{0.3}(Zn_{0.85}Mn_{0.15})_2As_2$ sample that you did using PDFGUI above. To run this code, you should have downloaded the script in a directory called DIFFPY-CMI, with a parent directory called **solutions** and another directory called **data** located at the same directory level as **solutions**. The **data** directory should contain the files **BaZn2As2_K-Mn-doped_300K.gr** and **BaZn2As2.cif**. Have a look at **fitCrystalGen.py** and the comments therein. We will explain numbered comments representing significant deviations from the previous DIFFPY-CMI examples.

 # 1-7: We begin as we have in previous examples, with necessary package imports, setting up some global parameters like the name of our fit, the location of the data and structure file, details about the measurement, and initial values for some fitted parameters. Importantly, we now need to provide initial values for the site occupancy as our sample is doped with Mn and K.

 # 8-9: We create a `make_recipe` function and load in our structure file, which does not contain Mn or K. (In this example we use the unedited CIF file from the original download and show how to make the modifications to the data directly in the script.)

 # 10: Since we've loaded in a structure of undoped $BaZn_2As_2$, we need to modify it to include our Mn and K dopants, just as in the PDFGUI example in this chapter. We can do this by looping over every atom in the structure and adding an Mn atom when we find a Zn or a K atom when we find a Ba. We use the coordinates of the found atom for each new atom we add, so that they are co-located in the structure.

11-20: Adopting the same path as in previous examples, we create a profile generator and a fit contribution, and give the latter our experimental profile and our PDF profile generator. We define our fit equation, create a fit recipe, give it our fit contribution, initialize values defining our experiment, and add a scale parameter to our fit recipe. We also add some space group symmetry constraints and a correlated motion parameter (δ_1).

 # 21: We have modified our loaded structure to include our Mn and K dopants, but we also would like to refine the occupancy of the doped sites. We'll need to create two new parameters for this, one for each occupancy: `Mn_occ` and `K_occ`. For convenience, we give them both the same tag, `occs`, so that we can free them both at the same time if we want to by using the tag. They can still be freed separately by using the variable names.

 # 22: So, we've created our occupancy parameters, but right now they don't truly *do* anything. If our fit varies them, they will not affect the fit residual. We need to tell the fit recipe to bind them to the occupancy of the sites of interest (Ba/K and Zn/Mn sites). We do this using constraint equations, through the `constrain()` function. We loop over every atom in the atomic structure associated with our fit recipe, and if the `label` attribute matches a pattern, we constrain the `occupancy` attribute. This shows tricks for how to automate repetitive tasks in the script. Check and double-check that the automation code is working correctly, but once it is verified, it is likely to actually reduce errors coming from copy-paste mistakes, for example.

The `constrain()` function is quite flexible. The first argument is the object we would like to constrain, and the second argument is a string representation of a parameter, or even an expression involving many parameters. This string can only contain parameter names which are already known to the fit recipe, i.e. user-defined or added to the recipe by a helper function. In our example, we use `Mn_occ` and `K_occ`, which we have already added to our fit recipe.

23-31: The remainder of our code proceeds identically to the previous examples. We only need to remember to include our new `occs` tag in our list of parameters to refine.

7

Investigating the tetragonal-to-orthorhombic phase transition in SrFe₂As₂

Benjamin A. Frandsen

7.1 Introduction and overview

This chapter investigates a structural phase transition in a crystalline material by performing structural refinements at a series of temperatures spanning the transition. We will use $SrFe_2As_2$ as an example case. We perform temperature-series refinements and inspect the temperature dependence of the fit results to glean information about structural phase transitions.

7.2 The question

$SrFe_2As_2$ is the parent compound for an interesting family of iron-based superconductors (Taddei *et al.*, 2016). For this chapter, the most important thing to know about $SrFe_2As_2$ is that it undergoes a long-range structural phase transition from a tetragonal structure (in the $I4/mmm$ space group) at high temperature to an orthorhombic structure (space group $Fmmm$) at low temperature. The orthorhombic structure distorts the square lattice into a rectangular lattice, reducing the rotational symmetry from four-fold in the tetragonal structure to two-fold in the orthorhombic structure. The crystal structure and orthorhombic distortion are illustrated in Figure 7.1. We would like to use PDF to understand this phase transition in more detail.

Several scientific questions are relevant to the structural phase transition in $SrFe_2As_2$, including which structural parameters change at the transition, the magnitude of the distortion, whether the short-range structure shows local symmetry breaking above the long-range phase transition, and how the behaviour of the local

Atomic Pair Distribution Function Analysis. Benjamin A. Frandsen, Oxford University Press.
© Simon J. L. Billinge and Kirsten M. Ø. Jensen (2023). DOI: 10.1093/oso/9780198885801.003.0007

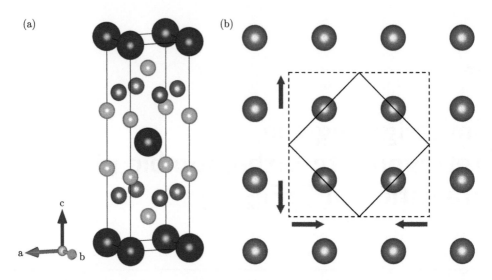

Figure 7.1 (a) Unit cell for the tetragonal structure of $SrFe_2As_2$. Purple, brown, and green spheres represent Sr, Fe, and As atoms, respectively. (b) View of one of the planes of iron atoms, with the projection of the tetragonal unit cell shown by the solid black lines and the orthorhombic unit cell by the dashed black lines. The thick arrows represent the orthorhombic distortion, turning the square lattice of iron atoms into a rectangular lattice.

and average structures across the phase transition may be related to the magnetic and superconducting properties of $SrFe_2As_2$-type compounds. However, in this chapter, we will limit our focus to determining whether or not we can detect the phase transition in the PDF data and, if so, if we can obtain a quantitatively accurate model of the low-temperature structure.

7.3 The result

From our PDF analysis, we confirmed that the high-temperature tetragonal structure undergoes a transition around 200 K. Moreover, we found that the low-temperature PDF data can be accurately modelled using the orthorhombic $Fmmm$ structure expected from earlier diffraction experiments. These results are discussed more fully in Frandsen *et al.* (2017, 2018).

7.4 The experiment

The experiment was performed on a powder sample of $SrFe_2As_2$ under the conditions given in Table 7.1. The files to download are listed in Table 7.2.

Table 7.1 Experimental details.

Facility	NSLS-II at Brookhaven National Laboratory
Beamline	28-ID-2
Date collected	October 2017
Detector type	Perkin Elmer amorphous silicon 2D detector
Sample geometry	Powder in 1 mm ID Kapton capillary
Sample environment	Cryostat with variable temperature from 150–250 K
Total exposure time	30 s
Wavelength	0.1867 Å
Q_{max}	25.0 Å^{-1}
Q_{damp}, Q_{broad} from calibration	0.0349 Å^{-1}, 0.0176 Å^{-1}

Table 7.2 Files to download.

Filename	Note
`SrFe2As2_*K.gr`	experimental PDF data at various temperatures
`SrFe2As2_tetragonal.cif`	CIF file for tetragonal structure
`SrFe2As2_orthorhombic.cif`	CIF file for orthorhombic structure

7.5 What next?

1. Download the files.
2. Get a `CIF` file from PDFITC to use as a starting point.
3. Create a PDFGUI project, and start a new fit for the tetragonal model.
4. Create the structure for the fit by loading the tetragonal CIF, load the data file for 246 K, and set the fit range to $1.5 \leq r \leq 50.0$ Å.
5. Fix the instrumental parameters to appropriate values, and set the refinable parameters to be variables, including the overall scale factor, lattice parameters, atomic positions, ADPs, and δ_1 or δ_2. You can use the space group information to load the appropriate atomic positions and ADPs automatically.
6. Perform a calculation with the starting structure to compare with the data and make sure it is at a good enough starting point. Use this to estimate a sensible starting scale factor.
7. Refine your model to obtain the best fit. Begin by only refining the parameters with the largest influence on the PDF (e.g. scale and lattice parameters), and

then slowly free more variables in the model. Inspect the fit (you should be doing this along the way), and verify that it is of reasonable quality, with R_w around 12%.

8. Once you are satisfied with the fit, run a temperature series macro fit, starting from the 246 K data set and continuing through to the 150 K data set.

9. Plot the temperature dependence of various fitting parameters, and look for anomalies suggesting a structural phase transition.

10. Set up a new fit using the orthorhombic structural model and the 150 K data set to determine if the orthorhombic model provides an accurate description of the low-temperature structure.

7.6   Wait, what? How do I do that?

7.6.1   Download the files

Follow the instructions in Section 3.6.1.

7.6.2   Create a PDFGUI project and start a new fit

Open up PDFGUI and create a new fit. See Section 3.6.3 if you need a refresher on how to do this in PDFGUI. Give your fit a descriptive name and save the project for safekeeping. We recommend frequent saves, which gives quicker recovery from crashes.

7.6.3   Get a CIF file from PDFITC to use as a starting point

Using STRUCTUREMINING in PDFITC is a great way to get CIF files to make a start on your refinement campaign. Please see Section 6.6.2 for more details. For convenience, we also provide the CIFs you need with the download files.

7.6.4   Load the structure and data files into your fit

Referring to Section 3.6.4 if necessary, load the tetragonal structure from the corresponding CIF. Verify that the phase information has been populated correctly; you can make any necessary changes directly within PDFGUI. Add the 246 K dataset in the same way as outlined in Section 3.6.3, and verify that it has read the correct scatterer type (x-rays) and $Q_{\max}$ value in the **Configure** panel. Change the fit range to be from 1.5 to 50 Å.

7.6.5   Fix the instrumental parameters to appropriate values
and set the refinable parameters

In the dataset configuration tab, set the appropriate values for Q_{damp} and Q_{broad} (listed in Table 7.1). Set refinable parameters for the scale factor, lattice parameters (recalling that $a = b$ for tetragonal symmetry), atomic positions, ADPs, and δ_1 or δ_2. If necessary, consult Section 6.6.6 for a refresher on how to do this.

7.6.6   Perform a starting calculation

Following the procedure outlined in Section 6.6.7, perform a starting calculation to ensure there is reasonable agreement between the model and the data. You can also compare the magnitudes of the calculated and observed PDF patterns to come up with a starting guess for the overall scale factor. In this case, we notice that the magnitude of the calculated PDF is about 1.5 times larger than that of the data, so we can set the initial data scale factor to 0.67.

7.6.7   Refine the tetragonal model

You are now ready to refine the model. If necessary, refer back to Section 3.6.7 for a reminder about how to run refinements in PDFGUI. We will refine the parameters sequentially, starting with the parameters that have the biggest effect on the calculated PDF and adding the others roughly in order of decreasing importance for the fit (see Section 6.6.8). If you need a hint for a good refinement sequence, see the Solution section (and see Section 6.6.8). You may also try either δ_1 and δ_2 to see which one gives a better fit. Eventually, you should reach a value of R_w around 12%.

7.6.8   Create and execute a temperature series macro based on your tetragonal fit

Now that you have a good fit at 246 K, we want to do the fits for all the other data sets collected at different temperatures. This can be done automatically within PDFGUI. Select your fit from the fit tree (this will serve as the template fit), then select Fits → Macros → Temperature Series. The menu will now allow you to select all the data files you want to include in your temperature series fit. Navigate to your data folder and highlight all data files (17 in total, ranging from 150 K to 246 K in steps of 6 K). They should now appear in the list of data sets, and, if possible, the temperatures are automatically extracted from the file names by PDFGUI. Please see the PDFGUI documentation for naming rules that accomplish this. We want to do the fits in descending temperature order, so select the Temperature tab at the top of the list until it orders itself appropriately. Then click OK, and you will see the fit tree become automatically populated with one fit for each temperature. Click on one of these automatically generated fits, and you will see that the initial value for each parameter is set to the refined value from the previous fit, which helps the fits start out reasonably close to the optimal structural model. To run the fits sequentially, select each of the fits (left click on one of the fits in the fit tree and then use CTRL+SHIFT+A to highlight all of them simultaneously) and then press the blue gear icon. It may take a few minutes for the program to run through each of the refinements sequentially.

7.6.9   Plot the temperature dependence of the tetragonal fit results

Once the fits are completed, we want to see how the results vary as a function of temperature to determine whether we have evidence of a structural phase transition occurring. To plot the temperature dependence of a refined parameter or R_w, select one of the fits and press CTRL+SHIFT+A to highlight all the others. From the Plot Control menu, select temperature for the X variable and whichever fitting parameter you want to visualize for the Y variable. You may display multiple fitting parameters on the same plot by holding down CTRL and selecting all the desired parameters from the Y menu, and then clicking the plot icon. An alternative way to view the temperature dependence of structural parameters is to select the structure under one of your fits in the fit tree, then use CTRL+SHIFT+A to highlight all the other structures, and then choose temperature for X and the desired structural parameter(s) for Y. Note that R_w is only available if you select the fits instead of the structures. Explore the temperature dependence of the fit results, and convince yourself that a significant change to the structure of $SrFe_2As_2$ occurs roughly in the middle of the temperature range.

7.6.10   Start a new fit using the orthorhombic model

From conventional diffraction analysis, we expect the low-temperature structure of $SrFe_2As_2$ to be orthorhombic. Here, we want to see if the expected orthorhombic structure describes the low-temperature PDF data well (or better than the tetragonal model, at least). Repeat the procedures in Sections 7.6.2, 7.6.4, 7.6.5, 7.6.6, and 7.6.7, but use the orthorhombic model instead of the tetragonal model and the 150 K data instead of the 246 K data. Due to the orthorhombic symmetry, the a and b lattice constants need not be equal to each other, so you may assign them to separate variables.

7.7   Problems

1. What happens to R_w for the tetragonal model below about 200 K? What are the physical implications of this?
2. How do the ADPs for the tetragonal model behave below 200 K? Why do they show this behaviour?
3. What other parameters show changes around 200 K, hinting at a change in the structure?
4. Compare R_w for the tetragonal and orthorhombic models at 150 K. What does this indicate about the low-temperature structure of $SrFe_2As_2$?

7.8   Solution

In Figure 7.2, we display the best fit for $SrFe_2As_2$ at 246 K using the tetragonal model. To arrive at this fit, we refined the parameters in the following order: we changed scale factor and lattice parameters in the first iteration; we then added the z-position of the arsenic sites in the second iteration; ADPs were adjusted in the third iteration; and δ_2 in the final iteration (chosen because it provided a slightly better fit than δ_1). After each iteration, we copied the refined parameter values to the initial values. The refined parameters are provided in Table 7.3.

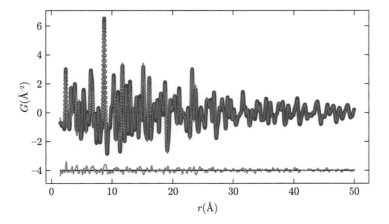

Figure 7.2 Fit to the experimental PDF data collected from SrFe$_2$As$_2$ at 246 K using the tetragonal structural model. Blue circles represent the data, the red curve the fit. The fit residual is shown by the green curve, offset vertically for clarity.

Table 7.3 Refined structural parameters of SrFe$_2$As$_2$ at 246 K from fits using the tetragonal model.

a (Å)	3.92717
c (Å)	12.3364
z for As (fractional)	0.63951 (equivalent to 0.36049, 0.13951, 0.86049)
U_{11}, U_{33} for Sr (Å^2)	0.00751, 0.00941
U_{11}, U_{33} for Fe (Å^2)	0.00893, 0.01133
U_{11}, U_{33} for As (Å^2)	0.00850, 0.00615
δ_2 (Å^2)	3.8624
Scale factor	0.6385
R_w	0.1197

These results are in general agreement with published results from Rietveld refinement. With $R_w = 0.1197$, the PDF fit obtained here is adequate, although perhaps not excellent. This is due to difficulties in making very-high-purity powder samples of SrFe$_2$As$_2$. Nevertheless, this is sufficient to serve as a template fit for the temperature series macro.

Key results from the temperature series refinements of the tetragonal model are shown in Figure 7.3. In panel (a), we plot R_w as a function of temperature, which displays a sharp rise starting between 192 K and 198 K, indicating that the tetragonal model provides an increasingly poor description of the material below this temperature. Panel (b) displays the in-plane ADP for the three different atom

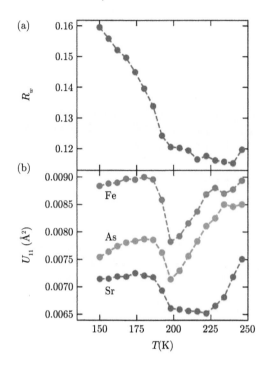

Figure 7.3 (a) Temperature dependence of R_w (a) and the U_{11} ADP for each atom type (b) determined from fits using the tetragonal structural model.

types, likewise showing a sharp rise below 198 K. ADPs capture the effect of thermal motion, so normally, we would expect smaller ADP values as the temperature is lowered. However, structural refinements can often produce anomalously large ADP values when an incorrect structural model is used, since the effect of the larger ADPs is to broaden the calculated PDF peaks, which may help capture the broadening or splitting of PDF peaks caused by an actual change in the symmetry of the structure. In this case, the ADPs increase sharply below 198 K because the tetragonal model is no longer the correct model to describe the structure. Together, these results demonstrate that the expected structural phase transition in $SrFe_2As_2$ is clearly observable in the PDF data and occurs between 192 K and 198 K.

Naturally, we would like to find a structural model that provides a better fit to the low-temperature PDF data, so we use the orthorhombic structure reported in the literature. The refined parameters for the orthorhombic model using the data collected at 150 K are provided in Table 7.4.

R_w for the orthorhombic model at 150 K is 0.1119, significantly better than that of the tetragonal model at 150 K (0.1596) and slightly better than that of the tetragonal model at 246 K. This indicates that the orthorhombic *Fmmm* model is appropriate to use for the low-temperature data. In other words, we have answered the scientific question posed at the beginning of this chapter and found a

Table 7.4 Refined structural parameters of $SrFe_2As_2$ at 150 K from fits using the orthorhombic model.

a (Å)	5.5801
b (Å)	5.5230
c (Å)	12.3014
z for As (fractional)	0.63963 (equivalent to 0.36037, 0.13963, 0.86037)
U_{11}, U_{22}, U_{33} for Sr (Å²)	0.00608, 0.00469, 0.00806
U_{11}, U_{22}, U_{33} for Fe (Å²)	0.00644, 0.00516, 0.00846
U_{11}, U_{22}, U_{33} for As (Å²)	0.00514, 0.00586, 0.00508
δ_2 (Å²)	3.5273
Scale factor	0.6256
R_w	0.1119

quantitatively accurate model of the structure at low temperature. As a next step, we could examine the temperature dependence of the refined orthorhombic model by conducting an additional temperature series macro. We could also perform a careful r-dependent analysis to determine if the orthorhombic distortion persists on short length scales even above the long-range structural transition temperature. See Chapter 9 for an example.

This concludes the current chapter. Congratulations! You now have experience investigating structural phase transitions through PDF analysis. Being able to perform sequential PDF refinements like those done here is also very useful for many other purposes, i.e. in analysis of data from *in situ* and *operando* experiments, and we hope it serves you well.

7.9 DIFFPY-CMI Solution

This section describes the DIFFPY-CMI solution. It may be skipped on a first reading, but it will be helpful to you when, in the future, you want to learn how to use DIFFPY-CMI. To get started with DIFFPY-CMI, follow the instructions in Section 3.9. The DIFFPY-CMI code for the solution will be in a Python file in the diffpy-cmi folder under solutions for this chapter in your downloaded data. Read through the comments there. They are expanded on below in greater detail where needed.

7.9.1 fitTSeries.py

This file presents a DIFFPY-CMI solution to the refinement of a series of temperature-resolved PDF data collected from the $SrFe_2As_2$ sample, testing two different structure models, just as you did using PDFGUI above. To run this code, you should have downloaded the script in a directory called Diffpy-CMI, with a parent directory called **solutions** and another directory called **data** located at the same directory level as **solutions**. The **data** directory should contain the files SrFe2As2_orthorhombic.cif and SrFe2As2_tetragonal.cif, as well and 17 PDF data files, with the general form SrFe2As2_$\langle$TTT$\rangle$K.gr, where $\langle$TTT$\rangle$ represents temperatures from 150 K to 246 K. This example will take some time to run, as we are running many PDF fits. Have a look at fitTSeries.py and the comments therein. We will explain numbered comments representing significant deviations from the previous Diffpy-CMI examples.

 # 1-21: The beginning of this example proceeds nearly identically to the simplest case, outlined in Chapter 3, with the exception that we do not refine Q_{damp} or Q_{broad}. We will not elaborate on each point here, but we define some details of our experiment, choose initial values for our refined parameters, and create a make_recipe function to build our recipe.

 # 22: It is in our **main** function where we start to see something new. We wish to fit multiple PDF datasets, each taken at a different temperature. We need to find each of these files, and it's a bit inconvenient to hard code the names of 17 different files into our script.

This is where the power of Python comes in. We have defined **DPATH** as a **Path**-type object. **Path** objects have many built in functions which are useful for file operations and are operating system-independent. We can use the **glob** function to search through all the files in **DPATH** and give us a list of those matching the argument of **glob**, a pattern of our choosing.

Our search pattern is f"*{GR_NAME_BASE}*.gr". Let's carefully dissect this. This is what we call an f-string, denoted by the leading **f** preceding the first quotation mark. This means that for any text nested in curly braces {}, Python will execute that text as code, and the text will be replaced by the value stored in that variable, or whatever is returned by executing the expression. The expression must return a quantity that Python can interpret as a string. **GR_NAME_BASE** contains the string "SrFe2As2_", which we know is contained in all the PDF files we'd like to find. The ∗ is a wildcard character, meaning it will match any character or series of characters. Finally, we know all our files end with .gr, so our search string is terminated with this.

 # 23: Now that we've got a list of PDF files to fit, we should probably try to associate each of these with a temperature. This is what PDFGUI does automatically. Here we show how to develop the code for doing this in Python that you can adapt for different situations. In this case, the measurement temperature has been written into the filename. For example, we can understand that SrFe2As2_228K.gr was measured at 228 K.

For parsing out the temperature, we will use a Python trick called "list comprehension", but you could also just as well use a **for** loop. First, to get the filename without its extension, .gr, we want to remove the .gr. We can do this using the built in `Path` function `stem`, which will return a string representation of the filename without the extension. Next, we can note that the temperature always immediately follows the single underscore _ in the filename and is always followed by K. We split the filename string into a list of strings using the _ delimiter passed to the `split` function. We take the last element of this list, then use the `strip` function to remove the trailing K.

Now, we have a list of strings representing the temperature each PDF was measured at. String representations of numbers can occasionally cause problems, so we want to convert these to something our code will recognize as numbers. We again use list comprehension for this, calling `int` on every entry of our temperature list to convert each to an integer.

24: Now that we've parsed out a list of temperatures from our data files, we might be tempted to begin fitting. We should think again, though, as both these lists are currently sorted based on the result of our `glob` call. `glob` makes a call to our operating system, and file sorting can vary between operating systems. We certainly don't want to obtain different results depending on which operating system we are using! So, here we sort the list of data files and temperatures together, in order of decreasing temperature. Recall that we converted our parsed temperatures from strings to integers, which facilitates this sorting.

 # 25: Remember, we are testing two different structures against this PDF temperature series. While we could hard-code the full path to both these structure files, we can also again make use of our `glob` function to find all the relevant files. This could be useful if we wanted to test several different structures.

 # 26: We've got a list of the relevant structure files now, so let's loop on each of them. We put this loop outside any loop on temperature, such that we can re-use each recipe for subsequent temperature points.

 # 27: We would like to have some awareness within our code of just which structure we are currently working with. To do this, we load in the structure file and get a string that represents the structure. Specifically, we get the short space group name, but we need to be sure we don't include any problem characters in this string, as we will be using it to write file names. We replace any "/" with "_on_", as the "/" character can cause issues with naming.

 # 28: We call our `make_recipe` function, prior to entering any loop on temperature. We use the first data file in our list here, but we could use any data file. We will be replacing this experimental profile later, once we enter our loop on all temperature points.

 # 29: We now begin our loop on each temperature point and its associated PDF data file. Keep in mind that this is nested inside our loop on all structure files.

 # 30: We want each fit for every temperature point and every structure to have a unique name. To achieve this, we automate the name creation. We include the chemistry, the structure type (space group), and the temperature.

 # 31: We need to tell our recipe about our PDF data at each temperature point. We create a new experimental profile and assign it into our recipe, and we tell this profile what r-range we want to fit over.

 # 32-36: The remainder of our code proceeds identically to the previous examples. We optimize our fit recipe by freeing parameters sequentially, and then we write results using our bespoke fit name. Each time the inner loop on our temperature and PDF file lists rolls over, the recipe will retain the best-fit values from the previous temperature. This is convenient as we would expect only small changes between temperature points.

Once this has finished, your directory will be populated with plots of the best fit at each temperature, for each structure, as well as text files describing the fitted parameters and fit statistics. These files can be post-processed and parsed using Python to generate plots of parameters as a function of temperature, or this could be done right inside our script with a few modifications. We leave this to the interested and motivated!

8

Simple modelling of nanoparticles: Size-dependent structure, defects, and morphology of quantum dot nanoparticles

Soham Banerjee and Kirsten M. Ø. Jensen

8.1   Introduction and overview

In the previous chapters, we have used PDF to characterize the structure of a range of bulk materials. Here, we will move to nanomaterials, something we already touched upon in Chapter 3 where we fitted a PDF obtained from platinum nanoparticles. Over recent years, characterization of atomic structure in nanoparticles has become one of the main uses of PDF, and here we will give an example concerning the atomic structure in CdSe nanoparticles. Unlike bulk materials, small nanoparticles do not give rise to sharp Bragg peaks but instead broad, diffuse scattering features. Quite often, we cannot interpret these using conventional crystallographic methods such as Rietveld refinement, but PDF is an excellent tool for studies of atomic structure in nanoparticles (Lindahl Christiansen *et al.*, 2020a).

As we have seen so far, the first step in PDF modelling is usually to find a good starting model that can be refined to fit to data from the sample in question. For most crystalline materials, such starting models can be found as CIFs in the millions of published crystal structures that are available in the literature and in structure databases. However, when analysing the structure of small nanoparticles, this is not always enough, as the size-induced structural differences may be so large that we cannot just assume that known crystal structures for the corresponding bulk materials can be fitted to the data. Nanomaterials often contain many structural defects and can have a distribution of particle sizes, and it can be challenging to describe such systems with a single crystal structure model. In Chapter 11 we show how DIFFPY-CMI can be used to develop bespoke models for discrete clusters, but here we show how to take the approach of massaging models

Atomic Pair Distribution Function Analysis. Soham Banerjee and Kirsten M. Ø. Jensen, Oxford University Press.
© Simon J. L. Billinge and Kirsten M. Ø. Jensen (2023). DOI: 10.1093/oso/9780198885801.003.0008

built from periodic structures to get the most information we can from the data without resorting to time-consuming discrete particle modelling. We demonstrate simple ways to model these rather complex problems while still extracting most of the information we are after. This requires some creativity in designing models and modelling approaches using PDFGUI, and the best way of doing this is highly dependent on the system in question. This is a very time-efficient approach and is almost always the best starting point. Only if a specific scientific question cannot be answered this way is it necessary to explore more complex models. There are many good examples of creative approaches to such problems in the literature, and here we focus on an important system – CdSe quantum dots (Masadeh *et al.*, 2007; Yang *et al.*, 2013). CdSe quantum dots have been extensively studied due to their optical properties and applications in solar cells and lighting. Their properties are highly size-dependent, and recently chemists have developed synthesis methods allowing very high control of particle size. This makes it possible to characterize the relation between size and structure in detail (Masadeh *et al.*, 2007; Yang *et al.*, 2013).

As we will see in this chapter, bulk CdSe generally takes the wurtzite crystal structure. A cutout of the structure is illustrated in Figure 8.1. In this structure, the anions sit in a hexagonal closed packed (hcp) arrangement, while the cations are in the tetrahedral sites in the structure. This crystal structure is very closely related to the zinc blende structure, where the anions form a cubic closed packed structure (ccp) with the cations in the tetrahedral sites. We often describe the difference between the hcp and ccp structures in terms of atom layering: as shown in Figure 8.1, the atomic layers in the wurtzite (hcp) structure can be described as having ABABAB stacking, while the layers in the zinc blende (ccp) structure are stacked ABCABC. Because both ABABAB and ABCABC stackings result in fully connected and unstrained bonding networks, they are close in energy and only differentiated energetically by weak secondary effects. Therefore, layered structures like CdSe are prone to *stacking faults*, where the layering does not follow either ABABAB or ABCABC through the whole structure but may have other sequences. Stacking faults are prominent in CdSe nanoparticles, and we will here use PDF to characterize them.

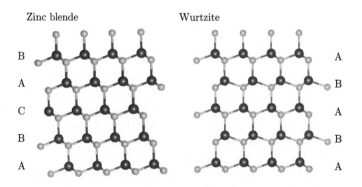

Figure 8.1 Cutouts of the zinc blende and wurtzite structures showing the ABCABC and ABABAB stacking. Cd is shown in pink and Se in green.

There will be quite a lot of steps to follow as we skip ahead to the Wait What? section (Section 8.6) and get a crib from what we have done in the PDFGUI project solution files.

8.2 The question

The goal of this study was to obtain quantitative structural information for CdSe quantum dot nanoparticles (NPs) (Masadeh *et al.*, 2007; Yang *et al.*, 2013).

For example, we wanted to know whether these small CdSe NPs have a predominantly wurtzite or zinc blende structure. Do they contain stacking faults? Can we obtain the crystallite size from the PDF data with reasonable accuracy? How does the atomic structure vary with NP size?

8.3 The result

Using PDF, we were able to characterize the structure and size of the nanoparticles and show that CdSe nanoparticles contain stacking faults. It was furthermore possible to quantify the stacking fault density using a two-phase wurtzite/zinc blende model. We could also detect a size-dependent homogeneous compressive strain that increased in magnitude quite dramatically with decreasing NP size. There was also a small but significant increase in the inhomogeneous strain in the form of an increased bond-length distribution of nearest neighbour bonds with decreasing size. This was all done using simple and quick PDFGUI fits. All of this is discussed in detail in Yang *et al.* (2013).

8.4 The experiment

X-ray total scattering data were collected from five different CdSe samples: all powders packed in Kapton capillaries. The experimental conditions are given in Table 8.1, and the files to download are listed in Table 8.2.

Table 8.1 Experimental conditions.

Facility	APS
Beamline	6-ID-D
Detector type	Mar345 2D image plate
Sample geometry	Powder in 1 mm ID Kapton capillary
Sample environment	Room temperature, ambient conditions
Q_{max}	19.0 Å^{-1}
Q_{damp}	0.058 Å^{-1}
Q_{broad}	0.0 Å^{-1}

Table 8.2 Files to download.

Filename	Note
CdSe-Bulk.gr	Bulk PDF data
CdSe-4.gr	Larg(er) NP data
CdSe-3.gr	Medium NP data
CdSe-2.gr	Small NP data
CdSe-1.gr	Ultra-small NP data
CdSe-Zincblende.cif	Zinc blende CIF
CdSe-Wurtzite.cif	Wurtzite CIF

8.5 What next?

1. Download the files listed in Table 8.2 into your working directory.
2. Get a CIF file from PDFITC to use as a starting point.
3. Plot the bulk CdSe PDF (`CdSe-Bulk.gr`) and the four CdSe nanoparticle PDFs (`CdSe-1:4.gr`) using a plotting program of your choice.
   (a) Estimate the structural coherence (spherical particle diameter) for the nanoparticle PDFs by visual inspection.
4. By fitting structural models to the PDF from the bulk sample, determine which candidate structure, wurtzite or zinc blende, best describes the data. For now, use isotropic ADPs for all atoms in your fits.
5. Determine if either structure (wurtzite or zinc blende) better describes the PDFs obtained from the nanoparticle samples. Use the spherical particle diameter ($sp_{diameter}$) to model the diameter of the particles.
6. Investigate if the application of anisotropic ADPs improves your refinements of the nanoparticle PDFs. It is enough to only consider the structure that best describes the bulk dataset from point 4.
   (a) Inspect the refined ADPs from your anisotropic model. Have any of these parameters refined to unexpected and non-physical values?
7. Construct a two-phase model with wurtzite and zinc blende that is able to do the following:
   (a) Improve the refinements compared to the single phase models described above.
   (b) Yield physical values for ADPs, in addition to the other structural parameters.
   (c) Extract an estimate for the stacking fault density from local structure refinements for both bulk and nanoparticle data.
8. Compare the refined values for the spherical particle diameter, ADPs, and overall agreement factors obtained in points 7, 6, and 4.

9. Compare the stacking fault densities between the five datasets.
10. Determine if the mixed phase model breaks down (e.g. yields non-physical results) for any of the datasets.

8.6 Wait, what? How do I do that?

8.6.1 Download the files

Follow the instructions in Section 3.6.1. You should now have five data sets, one PDF for bulk CdSe, and four separate CdSe nanoparticle (NP) PDFs.

8.6.2 Plot the PDF files to see what they look like

The data you just downloaded have been properly integrated, background/surfactant subtracted, normalized, and transformed to the PDF for you, as described in Chapter 4 and Appendix B. To get a flavour for the data and this procedure, see Figure 8.2, where the 2D detector images are shown along with the integrated 1D datasets $I(Q)$, the reduced structure functions $F(Q)$, and the $G(r)$ functions.

1. Now plot the five `.gr` files using a plotting program of your choice.
2. Inspect the raw PDFs in the r-range from 1.5–40 Å. Determine an approximate r-value (in Å) for which you think atomic-pair correlations terminate and become indistinguishable from noise ("random" oscillations). These r-values are our first estimates for the size of the nanoparticles. Make a note of your estimates for the four NP sizes; these will come in handy soon.

8.6.3 Get a `CIF` file from PDFITC to use as a starting point

Using STRUCTUREMINING in PDFITC is a great way to get `CIF` files to make a start on your refinement campaign. Please see Section 6.6.2 for more details. For convenience, we also provide CIFs with the download files.

8.6.4 Determine which candidate structure, wurtzite or zinc blende, best describes the bulk PDF data

Firstly, we want to characterize the structure of the *bulk* sample. You have two different structure candidates, zinc blende and wurtzite. In PDFGUI, start by setting up two fits for the bulk data set, one with the wurtzite structure and one with the zinc blende structure, using the CIFs provided. Name each of the two fits something you can remember; an easy choice is to name the fits based on the structure model and the dataset.

8.6.4.1 Fit the zinc blende model to the PDF from the bulk sample

We will start with the zinc blende model. The zinc blende structure is cubic with $F\bar{4}3m$ space group symmetry. Once you have imported your CIF, inspect the initial values for your structure parameters in the configure tab in PDFGUI.

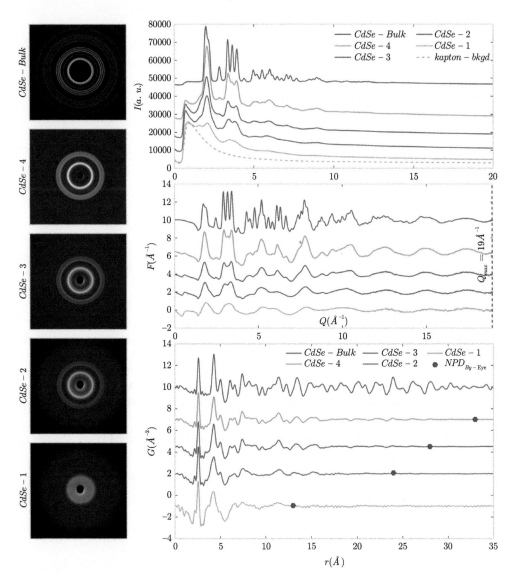

Figure 8.2 Data processing example and solution to NP diameter determination by visual inspection.

Make a note of the isotropic ADPs for each element that have been imported from the CIF. You may sometimes find CIFs where ADP values are included but are not reasonable guesses or are missing entirely. You are free to reset them to reasonable starting guesses, for example, 0.005–0.010 Å^{-2}, but make sure you include some value. In this case, the imported ADPs are reasonable starting values.

For this fit, we want to refine the cubic lattice parameter $(a = b = c)$, a scale factor, a peak sharpening parameter δ_2, and two isotropic ADPs, one for Cd and one for Se, where $U_{11} = U_{22} = U_{33}$ for each of the elements. This should give you five phase variables. Remember to set the instrumental parameters $(Q_{\text{damp}}$ and $Q_{\text{broad}})$ given in Table 8.1. Do this by navigating to the `Configure` tab for your `.gr` dataset and inputting the correct values. These should not be set as variables.

Before you start your refinements, confirm that you have provided reasonable starting guesses for each of the variables in the `Configure` tabs. Here is a cheat sheet to get you started. This is what you should see in the `Parameters` tab once the top of the fit node is selected in PDFGUI:

@1 : `a = b = c = 6.1` Å

@2 : `Scale Factor = 1.0`

@3 : `delta2 = 2.0` Å^2

@4 : `u11 = u22 = u33 = 0.005` Å^2 for Cd

@5 : `u11 = u22 = u33 = 0.005` Å^2 for Se

As mentioned in previous chapters, the @ tags are arbitrary in terms of the number you wish to assign. We could have chosen `@10` for the $a = b = c$ variable, `@101`, and `@102` for the U-values and so on. In fact there is a common strategy used to avoid accidentally reusing a parameter on two different tabs. For example, you could have a system where general parameters such as data scale factors are assigned to variables in the range 1–10, lattice parameters in the range 10–20, ADPs between 21 and 50, atomic positions between 50 and 100, and so on. Feel free to develop your own system and reuse it, you may find that will speed up your work and avoid mistakes, especially as your fits get more complicated.

For refinements of the bulk material, set the starting value of sp_{diameter} to zero. A value of zero for this field simply means that particle size effects are not included, i.e. the calculated PDFs are not damped by a spherical envelope function and are only attenuated due to the instrumental effects described by the Q_{damp} parameter.

Now, run your refinement, over an r-range of 1.5–40 Å.

You will see that the phase does not fit the PDF well, so we will have to try a different structure.

8.6.4.2   *Fit the wurtzite model to the PDF from the bulk sample*

We will now redo the procedure above but with the wurtzite model, which is a hexagonal structure with $P6_3mc$ space group symmetry. Start by assigning fit variables. Note that since wurtzite is hexagonal, you should have an additional phase variable compared to the zinc blende fit because $a=b\neq c$ in a hexagonal structure. This should give you a total of six phase variables. Run your refinement with the r-range 1.5–40 Å to ensure consistency. Our cheat sheet looks like this:

@1 : `a = b = 4.3` Å

@2 : `c = 7.0` Å

@3 : `Scale Factor = 1.0`

@4 : `delta2` $= 2.0$ Å^2
@5 : `u11` $=$ `u22` $=$ `u33` $= 0.005$ Å^2 for Cd
@6 : `u11` $=$ `u22` $=$ `u33` $= 0.005$ Å^2 for Se

8.6.4.3   Comparison of fit quality

Having fitted both the wurtzite and the zinc blende structures to the PDF from the bulk data set, you can now determine which model fits best by comparing both R_w values from the fit (given in the **Results** tab) and by visually comparing the two fits. We find that the wurtzite structure fits much better to the data than the zinc blende model. Note that this analysis could also have been done by fingerprinting the scattering pattern in reciprocal-space, as the bulk material gives rise to clear Bragg peaks, which can be indexed to the wurtzite structure. This is not the case for the smaller nanoparticles, which we can now start analysing.

8.6.5   Determine if either structure better describes the nanoparticle PDF data

For the bulk samples, it was clear that the wurtzite samples gave the best fit to the data. We will now test whether this is also the case for the nanoparticles, so we can set up refinements with each of the two phases for the four PDFs obtained from CdSe nanoparticles.

8.6.5.1   Fit the zinc blende model to the PDFs from CdSe nanoparticles

Again, we start with the zinc blende model, and we now want to set up fits for the four nanoparticle samples. You can do this in the PDFGUI project you are already working in; we do not have to start from scratch. In your existing PDFGUI project, create four fits, one for each of the `.gr` files from the nanoparticle samples. Import the appropriate `.gr` file into each of the fits, and set the Q_{damp} parameter from Table 8.1. Now, instead of importing the `.cif` again, you can simply copy the zinc blende structure from your bulk fit and paste it into your fits created for the nanoparticle samples. To do this, select (left click) the zinc blende phase in bulk fit in the fit tree panel, copy (`ctrl + C`), then click anywhere in the nanoparticle fit and paste (`ctrl + V`). All the parameters and constraints for this phase will appear in the new nanoparticle fit. Your fits are now set up with appropriate phase variables; we just need to add a new variable taking into account nanoparticle size.

As you already saw in Section 8.6.2, the PDFs from the nanoparticles truncated because of the small crystallite size. To take this into account in our modelling, we will introduce the sp_{diameter} (the spherical particle diameter), which we also used in Section 3.6.9. Therefore, add an additional variable for the sp_{diameter}. Set the initial guesses for this phase parameter (in the **Configure** tab) to your estimates after visual inspection of the nanoparticle PDFs.

For your four NP refinements, there should now be a total of six phase variables. As before, refine your nanoparticle PDFs over a wide r-range, from 1.5 to 40 Å. You should be careful and not necessarily refine all parameters at the same time initially. For example, it may be a good idea to initially keep the δ_2 value fixed while refining other variables, as was discussed in Section 6.8.

When you have better estimates for these variables, you can refine the δ_2 value along with other parameters. After your refinements of the bulk structures, you may also have better initial values for the ADPs.

8.6.5.2 Fit the wurtzite model to the PDFs from CdSe nanoparticles

You have probably already guessed it: the next step is to refine the wurtzite structure to the PDFs from the four nanoparticle samples. Again, you can use the wurtzite model that you have already set up by copying it into fits for each of the nanoparticle datasets, and add the sp_{diameter} to your model for a total of seven phase variables for the nanoparticle PDFs. Remember to set good starting values for the sp_{diameter}, and keep the Q_{damp} parameter fixed at the value given in Table 8.1.

8.6.5.3 Comparison of fit quality

We can now start comparing the results from the two models. Again, this is best done by simply looking at the fits and difference curves and by considering the R_w values obtained. For the largest particles, the wurtzite model fits slightly better to the PDF than the zinc blende model, but unlike the bulk materials, we find that there is little difference between the fit quality when comparing the two models. Also, the fits generally do not describe the PDFs all that well. Our fits and some of the results are shown in Figure 8.3, where we compare the R_w values and the refined sp_{diameter} from the two models. These are quite different between the models. This goes to show that if your fits are not good, your model is inadequate in some respect, and therefore care should be taken when ascribing scientific meaning to the values of your refined variables. See Section 5.8 for a discussion of uncertainties in refined parameters, but if the model is wrong or inadequate, you certainly cannot trust uncertainties coming from propagated statistical uncertainties because model errors are very likely to limit the precision of the parameter estimates. This is also true in other regression problems such as Rietveld refinement and is not specifically a PDF issue. In the next section, we will try to come up with better ways of describing the structure of the nanoparticles.

8.6.6 Application of anisotropic ADPs in your refinements using the wurtzite structure

So far in this chapter, we have only applied isotropic constraints for our ADPs. This is generally the way to go for most systems as the refinement of anisotropic ADPs requires very good models, high data quality, and a system whose symmetry allows this. Here, we are going to use anisotropic ADPs in a slightly different way than is often done in traditional crystallography. As discussed in Section 8.1, the difference between the zinc blende and the wurtzite structures is the stacking of atomic layers. In wurtzite this can be represented as ABABAB, while in zinc blende it is ABCABC, as illustrated in Figure 8.1.

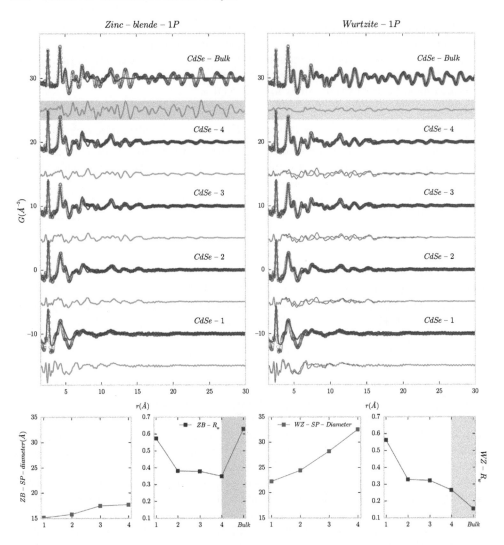

Figure 8.3 Top: Single phase fits with isotropic ADPs using the zinc blende model (left) and the wurtzite model (right). The blue curve shows the experimental PDF, the red curve the calculated PDF, and the green the difference between the two. In the fits using the wurtzite model (right), we have furthermore plotted the difference curve from the zinc blende fits as a red line. Notice that the two structures seem to describe different features of the PDF. Bottom: R_w values and sp_diameter obtained from the zinc blende (ZB) and wurtzite (WZ) fits.

The reason that the two structures gave fairly similar fits to the PDFs from nanoparticles is most likely the presence of stacking faults. In systems like this, one way of investigating this disorder is to allow anisotropic refinements of the ADPs. If the refined value of the ADP in the stacking direction is much larger than those perpendicular to it, it could indicate significant stacking disorder in the stacking direction. In our wurtzite model we are lucky: the stacking direction is along the crystallographic z-direction. We can therefore explore the presence of stacking faults by assigning U_{33} to a different variable than U_{11} and U_{22}.

If U_{33} refines to a much larger value than the $U_{11} = U_{22}$ variable, it is a strong indication for stacking disorder in the current case. In the zinc blende structure, the stacking direction is $\langle 111 \rangle$. It is still possible to do this same analysis in the zinc blende structure, but it is more complicated because the off-diagonal terms of the anisotropic displacement factors would have to be set correctly. This is beyond the scope of the current discussion.

We will now test the influence of applying anisotropic constraints in the wurtzite model. We will also include refinement of the z-positions of the atoms to account for any displacement of the atoms from the starting coordinates. We will only refine the z coordinates as changing the value for the rest of the atomic fractional coordinates would break space group symmetry, and in this case we don't want to do that. Local symmetry breaking will be explored in Chapter 9.

1. In all of your refinements so far, you have varied your ADPs per element by setting $U_{11} = U_{22} = U_{33}$ for Cd and $U_{11} = U_{22} = U_{33}$ for Se, respectively. To test the influence of anisotropic constraints for the hexagonal wurtzite structure, simply add two additional variables to your refinements by setting new variables for U_{33} for Cd and Se.

2. Refine the fractional position of Cd and Se along the c axis. The best way of doing this is by using the "Symmetry constraints" option available in the Constraints tab in PDFGUI by choosing "Constrain positions", as described in Section 6.6. This will automatically make sure that only crystallographic positions that are allowed to change by symmetry will be refined. If you need to confirm the initial guesses and variables, here is an example configuration template for the 4 nanoparticle datasets, with a total of 10 phase variables. You should by now be familiar with how refined parameters differ for bulk and nanoparticle PDFs.

@1 : a = b = 4.3 Å
@2 : c = 7.0 Å
@3 : Scale Factor = 1.0
@4 : delta2 = 2.0 Å^2
@5 : u11 = u22 = 0.005 Å^2 for Cd
@50 : u33 = 0.005 Å^2 for Cd
@6 : u11 = u22 = 0.005 Å^2 for Se
@60 : u33 = 0.005 Å^2 for Se
@7 : spdiameter Your estimate, or the value refined from isotropic wurtzite refinements
@8 : z = 0 for Se1
@9 : z = 0.375 for Cd1

Now run your refinements for the five PDFs with anisotropically constrained ADPs, and as before fix your refinement range to 1.5–40 Å.

Pay close attention to the fit quality and to the refined U_{33} values. You should see a slight improvement in fit quality for your nanoparticle samples and enlarged U_{33} values compared to U_{11} and U_{22} for either Cd or Se. You will most likely get values of the order of 0.15–0.20 Å^2, which is non-physically large (physical values for U due to thermal motion are 10–20 times smaller, typically in the range 0.005 to 0.02 Å^2). Such high refined U_{33} values are a clear indication of stacking disorder in these layered materials.

8.6.7 Mixed phase wurtzite/zinc blende refinements

Having established that the particles contain stacking disorder, we can now come up with a model that tests this hypothesis. To do this, we will construct a two-phase model (as we did in Chapter 5) that can extract a stacking fault density for both CdSe bulk and nanoparticle data. The idea is that the presence of stacking faults make the nanoparticle structure somewhere in between wurtzite and zinc blende. Both stacking sequences begin as AB, with wurtzite appending an A whereas zinc blende appends a C layer, as seen in Figure 8.1.

If we take wurtzite as the reference structure, and we define $P(C)$ as the probability of the third layer being a C with respect to the previous two layers defined as AB, we see that wurtzite is the case where $P(C) = 0$ and zinc blende is $P(C) = 1$. In the presence of stacking faults, $P(C)$ will be some number between 0 and 1, and the stacking fault density (with respect to wurtzite) will be given by $P(C)$.

We will set up a two-phase fit that allows determination of $P(C)$. Here, we will take advantage of a nice feature in PDF-fitting: we are free to choose which r-range we want to refine the model over. To obtain the stacking fault density $P(C)$ as described above, we need only refine over the r-range covering three stacking layers. In this way we are just counting, locally, the number of ABA vs. ABC stacking sequences. Because of the size of the CdSe$_4$ tetrahedra, this is roughly in the range 1–10 Å.

However, before we do that, we will do a refinement over the full r-range to obtain good values for the sp_{diameter} that we will then fix later for the low-r fits. Having a large r-range is critical for getting an accurate nanoparticle diameter.

We again want to set up five new fits, one for the bulk PDF and each of the nanoparticle PDFs. Just as before, we do not have to start from scratch: copy-paste your PDFGUI fits with any of the single-phase refinement routines (sticking to isotropic ADP models) so that you again get five new fits in your project. Now add the second phase to your fits, so that they contain both the zinc blende and wurtzite structures.

We will set independent variables for the cubic and hexagonal lattice parameters, as well as the fractional z coordinate for Cd and Se in the wurtzite phase. However, we fix the ADPs for Cd and Se to take the same value in each of the two phases.

The scale factors in the fits can be set up to directly reflect the stacking fault densities. As described in Section 5.6.4, the scale factors in two-phase fits can either be set up as two independent phase scale factors or with an overall dataset scale factor and a phase scale factor that expresses the ratio between the two phases. Here, we will choose the latter option. To set up the "overall" scale factor for the dataset, go to the Dataset `Constraints` tab. Here, you can set up your first scale factor variable as e.g. `@1`, but remember that you can choose the variable tag number as you like. We can now set up the scale factor for phases. Go to the wurtzite phase constraint tab, and set the scale factor as a new variable, e.g. `@2`.

You can set the initial value for this as 0.5. Now go the zinc blende phase. Here, you should set the scale factor as `1-@2`.

By doing this, we are able to directly use the phase scale factor as a measure for the stacking fault density. Our relative phase fraction constraint (@2) has an initial guess of 0.5, which assumes, as a starting point, a 1:1 ratio of wurtzite to zinc blende.

Here is a template, including initial guesses, for your two-phase nanoparticle refinements. We have marked which phase each `@` variable tag should be applied to, ZB for zinc blende; WZ for wurtzite. If a variable is flagged as being "equal" (ie. `WZ = ZB`), it implies that the parameter value should be the same between the two phases, that is, they must refine to equal values.

Dataset:@1 : `Scale Factor` $= 1$
      WZ:@2 : `Scale Factor` $= 0.5$
      ZB:1-@2 : `Scale Factor` $= 0.5$
      WZ:@3 : `a` $=$ `b` $= 4.29595$ Å
      WZ:@4 : `c` $= 7.01397$ Å
      ZB:@5 : `a` $=$ `b` $=$ `c` $= 6.1212$ Å
WZ $=$ ZB:@6 $=$ `delta2` $= 5.0$ Å^2
WZ $=$ ZB:@7 $=$ `spdiameter` Your estimate, or the value refined from anisotropic
                 wurtzite refinements
WZ $=$ ZB:@8 $=$ `u11` $=$ `u22` $=$ `u33` $= 0.01$ Å^2 for Cd
WZ $=$ ZB:@9 $=$ `u11` $=$ `u22` $=$ `u33` $= 0.01$ Å^2 for Se
      @10 $=$ `z` $= 0$ for Se1
      @11 $=$ `z` $= 0.375$ for Cd1

Start by doing this refinement over the full r-range, 1–40 Å, as we have done so far. When the fit is done, make sure to copy your refined parameters to be the new initial values in the parameter tab for the fit. We can now do the fit of the local range (1–10 Å) in order to determine the stacking fault density. In this fit, we do not want to fit the sp_{diameter} as a local range refinement will not yield a reliable value for this parameter. You should therefore fix it to the values obtained in the fits over the 1–40 Å range. Now, you can do the refinements over the local range for the three datasets and obtain stacking faults densities.

8.7 Problems

1. Now you can inspect your results. Do you have a satisfactory fit for all samples? Can you extract structural information even for the smallest nanoparticles?
2. We used the fits with anisotropic ADPs to investigate if there may be stacking faults in the particles, which we saw from the non-physical refined values of U_{33}. Are the refined ADP values in the two-phase fits physical?
3. What is the stacking fault density in the nanoparticles? Is there any effect from nanoparticle size? Compare to the results from the bulk sample.
4. Consider also the refined unit cell parameters. The most reliable values for this are obtained from fits over the full r-range of 1–40 Å. Do you see an effect of particle size?
5. We can also compare the PDFs directly; plot the four nanoparticle PDFs, and inspect what happens to the first peak. What do you see? What do the differences between the PDFs illustrate?

8.8 Solutions

In Figure 8.4, we compare our fits using the zinc blende (ZB) model, the wurtzite model (WZ), and the two-phase model (2P) to the `CdSe-4.gr` dataset. We see a large improvement by applying the two-phase model. Generally, the two-phase model provided a good agreement with the experimental PDFs obtained from the bulk materials and the three largest nanoparticle sizes. The fit to the PDF from the smallest nanoparticles is not as good, and we should therefore be careful not to extract too much information from this fit. It appears that the structure of the smallest nanoparticles may be quite different from the distorted wurtzite model as the peaks from the model (see the peaks around 6–7 Å) are not at the right positions. The structure of the smallest particles is discussed in more detail by Yang et al. (2013).

For the larger particles, the two-phase models give a good description of the experimental PDFs. We can therefore extract stacking fault densities of about 30% for the nanoparticle samples and 8% for the bulk materials. This agrees well with the initial one-phase fits we did for the nanoparticle samples, where there was not a clear difference between the fits using either the zinc blende or wurtzite structure – the high stacking fault density means that the structure is somewhere in between the two. The fit quality did not improve significantly between the wurtzite fit applying anisotropic ADPs and the two-phase fits, but the important point is that we can extract stacking fault densities and get physical values for the ADPs in the two-phase fit. We get ADP values around 0.02–0.05 Å^2, which is high but reasonable for strained nanoparticles, as discussed further below. It is important to note that the two-phase model does not mean that we have two distinct phases in the sample: we do not have some zinc blende and some wurtzite particles. Instead, the two-phase model means that the structure of the nanoparticles is somewhere in between that of zinc blende and wurtzite, and we can use the model to extract stacking fault densities from refinements to the local range.

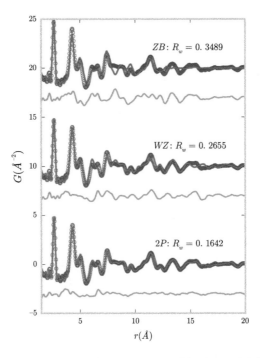

Figure 8.4 Fits to the CdSe-4 dataset using the zinc blende model (ZB), wurtzite model (WZ), and the two-phase model (2P). Both a visual inspection of the fits and the R_w values show a large improvement with implementation of the two-phase model.

When comparing the refined unit cell parameters obtained from the fits over the full r-range, the lattice constants of both the wurtzite and zinc blende phases decrease with decreasing particle size. This indicates a size-dependent compressive strain in the particles. We can actually see this directly in the experimental PDFs. If you compare the position of the first peak (try plotting the PDFs on top of each other), we directly observe that the peak shifts to shorter distances with smaller particle size. The peak also gets much broader with decreasing size. You can do a quantitative analysis of this if you fit the peak with, for example, a Gaussian function and extract peak position and width. The shift of the peak again represents a large homogeneous, compressive strain in the nanoparticles, while the peak broadening shows that the structure of the particles contains a large inhomogeneous strain. This increased strain with decreasing size is consistent with the change in structure seen for the smallest nanoparticles, where we could not obtain a good agreement between the model and the experimental PDF.

That is it for CdSe nanoparticles for now. We hope that this and other analyses of nanoparticle structure (Lindahl Christiansen *et al.*, 2020a) can be of inspiration for your own modelling. In the next chapters, we analyse the structure of other disordered materials.

8.9 DIFFPY-CMI **solution**

This section describes the DIFFPY-CMI solution. It may be skipped on a first reading, but it will be helpful to you when, in the future, you want to learn how to use DIFFPY-CMI. To get started with DIFFPY-CMI, follow the instructions in Section 3.9. The DIFFPY-CMI code for the solution will be in a Python file in the diffpy-cmi folder under solutions for this chapter in your downloaded data. Read through the comments there. They are expanded on below in greater detail where needed.

8.9.1 Running DIFFPY-CMI examples

See the instructions in Chapter 7 for help on running Python files.

8.9.2 `fitCdSeNP.py`

This file presents a DIFFPY-CMI solution to the refinement of a series of PDFs collected from the various CdSe quantum dot nanoparticle samples, testing various different structure models and fitting approaches, just as you did using PDFGUI above. To run this code, you should have downloaded the script in a directory called DIFFPY-CMI, with a parent directory called `solutions` and another directory called `data` located at the same directory level as `solutions`. The `data` directory should contain the files `CdSe-Wurtzite.cif` and `CdSe-Zincblende.cif`, as well as five PDF data files. This example will again take some time to run as we are running many PDF fits. Have a look at `fitCdSeNP.py` and the comments therein. We will explain numbered comments representing significant deviations from the previous DIFFPY-CMI examples.

 # 1-8: The opening of this script proceeds similarly to previous examples, and we will not elaborate on each point here. We define some details of our experiment and choose initial values for our refined parameters.

 # 9: Recall that in our PDFGUI example, we tested both one- and two-phase models against our PDF data. While we could likely write a generalized `make_recipe` function to handle an arbitrary number of phases, we will instead here simply write two functions, `make_recipe_one_phase` and `make_recipe_two_phase`, to handle one- and two-phase models, respectively. Remember, we also tested both isotropic and anisotropic ADP constraints, so we give `make_recipe_one_phase` a new argument, `adp_iso`, which will take a boolean, specifying if ADP terms are to be considered isotropic.

 # 10-18: This portion of our example proceeds just as in the second DIFFPY-CMI example in Chapter 7, where we looked at nanocrystalline Pt. We load in a structure and our data and create an experimental profile, fit contribution, and PDF profile generator. We define a fit equation, again using a spherical characteristic function, create a fit recipe, and add various fit parameters to the recipe.

19-23: This function needs to be able to handle two cases, creating a fit recipe with *either* isotropic *or* anisotropic constraints on all ADP parameters. To handle this, we will use an `if ... else` statement. If our new argument `adp_iso` is `True`, we will adopt one behavior; otherwise, another behavior.

If `adp_iso` is `True`, we use `constrainAsSpaceGroup` to generate a set of parameters to add to our fit recipe, but we only add those pertaining to the lattice and atomic positions to our fit; we do not add any ADP parameters generated using `constrainAsSpaceGroup`. Instead, we create two new fit parameters, `Cd_Uiso` and `Se_Uiso`, for constraining isotropic ADPs of Cd and Se atoms, respectively. We then loop over all atoms in the structure, constraining the `Uiso` attribute to either `Cd_Uiso` or `Se_Uiso`, depending on the value of the `element` attribute of the atom.

 # 24-26: If our new argument `adp_iso` is anything but `True`, we will allow for anisotropic constraints on all ADP parameters. This is done inside an `else` statement simply by adding ADP parameters generated by `constrainAsSpaceGroup` to our fit recipe, alongside lattice and atomic position parameters.

 # 27: Once we've done all this, our fit recipe can be returned from our function.

 # 28-48: Now we handle the two-phase case with a new function, `make_recipe_two_phase`. In this case we make use only of isotropic ADP constraints, so we have no need for a new argument. This function follows the same route as the two-phase `make_recipe` function in the DIFFPY-CMI section of Chapter 5. Notably, we include a single spherical characteristic function that describes PDF damping for both phases, together. We also create `Cd_Uiso` and `Se_Uiso` parameters and use each to constrain isotropic ADP parameters of Cd or Se atoms, respectively, across both structures. Similarly, we use one correlated motion parameter, δ_2, across both structures.

 # 49: Now we will construct our `main` function. We have several permutations of one- and two-phase fits to try with different constraints (isotropic and anisotropic) on our ADP parameters. We could hard code each of these individually, but we're going to leverage the power of Python to save ourselves some headache.

 # 50: Just as in the DIFFPY-CMI example in Chapter 7, we will use the `glob` function to find all PDF files in our data directory. We also want to give our script some awareness of just which file it is working with, but we certainly don't need to include the whole file name. Here we use list comprehension and some built-in `Path` functions to parse out an identifying string from each PDF datafile we find.

 # 51: Again as in the DIFFPY-CMI example in Chapter 7, we will use the `glob` function to find all structure files in our data directory and use list comprehension and some built-in `Path` functions to parse out an identifying string from each of these structure files.

 # 52:   We want to handle both isotropic and anisotropic ADP constraints, so we create a list of strings to keep track of which case we are working on. We can loop over this list later to be sure to consider all constraint types.

 # 53:   We chose for our outermost loop to be on all the PDF datafiles we have found, paired with their associated identifying strings. This will guarantee that we can apply all necessary fitting approaches to each dataset. If a given fitting approach is not appropriate for a given dataset, we can simply skip its application using `continue`.

 # 54:   We use the identifying string we have parsed out from our PDF files to build a bespoke name for our fit, and we include "two_phase" in our fit name. We will execute the two-phase fits outside any loop on the structure files, and just as in the PDFGUI example above, all our two-phase fits will use isotropic constraints on the ADP parameters.

 # 55-60:   In these steps, we proceed much as we have in previous DIFFPY-CMI examples. We call our function `make_recipe_two_phase`, and we give it both of our structure files, create a list of parameters and/or tags we want fitted, freeing them sequentially and fitting. Finally, we write our results to files using our bespoke fit name.

 # 61:   Much like the DIFFPY-CMI example in Chapter 7, we now loop over all the structure models we want to test, paired with the identifying string which we've parsed out from the structure file names.

 # 62:   We want to consider both isotropic an anisotropic constraints on the ADP parameters for each structure file and each PDF, so we loop on the list we created earlier, which contains strings representing the case we are working on, and we nest this loop inside that on the structure files and PDF files.

 # 63:   We will again create a bespoke name for our fit, this time using more information. We include the identifying string we have parsed out from our PDF file, the name of the structure we are testing, and the string identifying the type of symmetry constraints on our ADP parameters.

 # 64:   We call our single-phase make recipe function, `make_recipe_one_phase`, to build our recipe. We do this inside our loop on structure files, so we we test each structure file separately. We also construct a simple logic evaluation to decide which boolean to pass to our new `adp_symm` argument.

 # 65:   There are some situations where a given space group combined with a given atomic basis permits only isotropic ADP constraints, by symmetry. In such situations, the isotropic and anisotropic constraints on ADP parameters become

equivalent. We would like to avoid duplicating work, and save ourselves some time, so we get a bit creative and devise a test for this.

We free only the ADP parameters in our fit, and then check each parameter name. If all of the names contain the string "iso", we know that we are working with only isotropic ADP constraints. If, however, we see that we should be be working with anisotropic ADP constraints, we know we have encountered the special case where the two are equivalent. We tell our script to continue to the next iteration of the loop, and skip any fitting.

 # 66-70: Now we proceed just as in previous DIFFPY-CMI examples. We create a list of parameters and/or tags we want fitted, freeing them sequentially and fitting. Finally, we write our our results to files using our bespoke fit name.

Once this has finished, your directory will be populated with plots of the best fit for each different CdSe sample, using each of the relevant permutations of ADP constraints and structure pairings which you tested using PDFGUI earlier in this chapter. Have a look at these and compare them to your PDFGUI results!

9

Local structure in a crystal with short-range ordered lower-symmetry domains: Local iridium dimerization and triclinic distortions in cubic CuIr$_{1.76}$Cr$_{0.24}$S$_4$

Emil S. Bozin

9.1   Introduction and overview

This chapter introduces an r-dependent fitting methodology as a useful tool for estimating a relevant length scale, often also referred to as "structural coherence" or "correlation length", of the underlying short-range structural order. The PDF is a total scattering-based technique that encompasses both Bragg and diffuse scattering information. It provides structural information on continuous length scales from very short range (near and next nearest neighbours), through intermediate to long range (tens to hundreds of Ångstroms) depending on the reciprocal-space resolution of the measurement. Here we use CuIr$_{1.76}$Cr$_{0.24}$S$_4$ as an exemplar system to illustrate how to detect a local broken symmetry state and how to estimate the range or correlation of the local order. This is done using careful refinements of a high symmetry model to the experimental data. This is an example of "orbital degeneracy lifting" (Božin et $al.$, 2019) that results in spontaneous local symmetry breaking in transition metal systems and, more generally, polymorphous network behaviour (Zhao et $al.$, 2022).

9.2   The question

The spinel material CuIr$_2$S$_4$ displays a metal-to-insulator transition on cooling at ~230 K involving lattice, charge, orbital, and spin components of the system. It is accompanied by a lowering of the average crystal symmetry from a cubic ($Fd\bar{3}m$)

Atomic Pair Distribution Function Analysis. Emil S. Bozin, Oxford University Press.
© Simon J. L. Billinge and Kirsten M. Ø. Jensen (2023). DOI: 10.1093/oso/9780198885801.003.0009

high-temperature metallic state to a triclinic ($P\bar{1}$) low-temperature insulating state (Radaelli *et al.*, 2002).

At high temperature, Ir resides on an undistorted pyrochlore sublattice of $CuIr_2S_4$ comprising regular corner-shared tetrahedra. At low temperature, this sublattice distorts, with a fraction of its Ir ions getting closer to each other and forming dimers, which then form a long-range ordered pattern. This is illustrated in Figure 9.1, where the appearance of Ir–Ir dimers reveals itself in the PDF as a new peak corresponding to short Ir—Ir contacts. If some of the iridiums are chemically substituted by Cr, the metal-to-insulator average symmetry-lowering transition is rapidly suppressed, and the average structure remains cubic in $Fd\bar{3}m$ down to low temperature. However, curiously, dimerization also takes place in Cr-substituted derivatives, resulting in observable local distortions (Figure 9.1 (e)) at a temperature that is comparable to that in the parent $CuIr_2S_4$ but with long-range dimer order clearly absent (Božin *et al.*, 2014). We would like to utilize PDF analysis to explore the detection of this local dimer state and to determine its spatial extent in the $CuIr_{1.76}Cr_{0.24}S_4$ sample at 100 K.

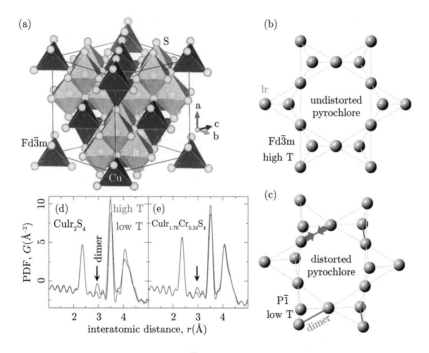

Figure 9.1 (a) High-symmetry cubic $Fd\bar{3}m$ structure of $CuIr_2S_4$ spinel. (b) Undistorted Ir pyrochlore sublattice of $CuIr_2S_4$ (high temperature). (c) Distorted Ir pyrochlore sublattice featuring Ir–Ir dimers (low temperature, $P\bar{1}$ symmetry). (d) Signature of Ir–Ir dimerization in experimental PDF of $CuIr_2S_4$. At low temperature, a new short Ir–Ir distance is observed as compared to high temperature. (e) Dimerization at low temperature in $CuIr_{1.76}Cr_{0.24}S_4$ has weaker yet clearly observable fingerprint resembling that observed in $CuIr_2S_4$ parent.

9.3   The result

From our PDF analysis we discovered that, despite the average crystal structure retaining its cubic $Fd\bar{3}m$ character at low temperature, the local structure of Cr-substituted $CuIr_2S_4$ spinel does distort on cooling, reflecting Ir dimerization below 180–200 K. Furthermore, from r-dependent PDF fits to the low-temperature data, we found that short-range ordered lower-symmetry domains of dimerized Ir existing within the nominally cubic $CuIr_2S_4$ matrix have spatial extents in the range of 1–2 nanometres. These results and the implications for the properties of the material are discussed in greater detail in Božin *et al.* (2014, 2019).

9.4   The experiment

The experiment was performed on a finely pulverized polycrystalline sample of $CuIr_{1.76}Cr_{0.24}S_4$ under the conditions given in Table 9.1. The files to download are listed in Table 9.2.

Table 9.1 Experimental details.

Facility	APS at Argonne National Laboratory
Beamline	11-ID-C
Date collected	Nov 2011
Detector type	Perkin Elmer amorphous silicon 2D detector
Sample geometry	Powder in 1 mm ID Kapton capillary
Sample environment	Cryostream with variable temperature from 90–300 K
Total exposure time	5 s
Wavelength	0.1257 Å
Q_{max}	27.0 Å^{-1}
Q_{damp}, Q_{broad} from calibration	0.038 Å^{-1}, 0.018 Å^{-1}

Table 9.2 Files to download.

Filename	Note
`CuIr1.76Cr0.24S4-q27r70t100-11IDC-APS.gr`	Experimental PDF data at 100 K
`Fd-3m.cif`	CIF file for cubic $CuIr_2S_4$ structure

9.5   What next?

The following is the crib for carrying out an r-dependent fitting on a single dataset.

1. Download the files.
2. Get a CIF file from PDFITC to use as a starting point.
3. Adapt the CIF file for $CuIr_2S_4$ to reflect Cr substitution.
   (a) Open up the CIF file in a text editor.
   (b) Modify the section with the atomic positions and occupancies to reflect the nominal Cr/Ir composition of the sample.
4. Create a PDFGUI project and start a new fit for the cubic model.
   (a) Create the structure for the fit by loading the cubic CIF.
   (b) Load the data file for $CuIr_{1.76}Cr_{0.24}S_4$ at 100 K.
   (c) Set the fit range to $1.5 \leq r \leq 50.0$ Å.
5. Set up the fit parameters.
   (a) Fix the instrumental parameters to appropriate values.
   (b) Set the refinable parameters to be variables, including the overall scale factor, lattice parameters, atomic positions, and ADPs. You can use the space group information to load the appropriate atomic positions and ADPs automatically.
   (c) Set the δ_1 parameter to a reasonable value (e.g. 2.0 Å) and fix it, as it should not be refined in this example.
6. Adjust the starting parameters and execute the fit.
   (a) Perform a calculation with the starting structure to compare with the data and make sure it is at a good enough starting point. Use this to estimate a sensible starting scale factor and lattice parameter.
   (b) Refine your model to obtain the best fit. Keep in mind that in this example it is not possible to obtain a good fit over the entire range, given that the local and average structures are *not* consistent. The fit will result in R_w around 25%, reflecting the inability of the model to fit short-range PDF data.
7. Examine the wide-range fit.
   (a) Plot the final wide-range fit.
   (b) Observe the misfit, and explore its length scale and character in the difference curve. This constitutes the discovery of the short-range broken symmetry state.
   (c) Familiarize yourself with the outcome of the wide-range fit, and plan for r-dependent fit.
8. Set up and execute variable-range fits.
   (a) Use the best fit parameter values found in the previous step as starting values. Save the PDFGUI project under a new name. Save your work frequently.

   (b) Set up and run an r-series macro fit on the 100 K dataset. Change the fit range by using a fixed fit minimum r-value and variable fit maximum r-value in 1 Å steps by starting from the 1.5–5 Å range and continuing through to the 1.5–50 Å range.

9. Evaluate the fit results to answer the scientific question.
   (a) Plot the fitting range dependency of various fitting parameters.
   (b) Look for anomalies suggesting a change in the structural response with change of interatomic distance.

9.6 Wait, what? How do I do that?

9.6.1 Download the files

Follow the instructions in Section 3.6.1.

9.6.2 Get a CIF file from PDFITC to use as a starting point

Using STRUCTUREMINING in PDFITC is a great way to get CIF files to make a start on your refinement campaign. Please see Section 6.6.2 for more details. We also provide a structure model with the download files.

9.6.3 Adapt the CIF file

As was done in Chapter 3, we will also here use a CIF to set up the structure of $CuIr_{1.76}Cr_{0.24}S_4$. Prior to using the CIF in PDFGUI, it is first necessary to adjust it to the correct nominal composition of the sample, since the downloaded CIF file corresponds to the structure of the $CuIr_2S_4$ parent system, and the sample of interest here contains 12% of Cr substituting for Ir. The original CIF therefore has to be adapted to reflect the change in chemical composition of the sample (see Section 6.6.3). In the CIF file provided with the downloaded data, there are three adjustments that need to be made. First, in the loop describing oxidation number spanning lines 241–246 in the CIF file, Cr should be added as Cr3+ 3.

   Second, in the loop describing atomic sites spanning lines 247–260 in the CIF file, Cr information should be added as Cr1 Cr3+ after Ir1, reproducing all Ir coordinates and ADPs, since Cr and Ir share the same site. Finally, in the same loop the occupancies of Ir and Cr need to be adjusted appropriately to reflect the nominal sample stoichiometry. Occupancy is described in the next-to-last field of the line describing atomic sites, and typically has a value of 1. The Ir occupancy should be set to 0.88 and Cr occupancy to 0.12. Save the CIF file under a unique name.

9.6.4 Create a PDFGUI project and start a new fit

Open up PDFGUI and create a new fit. Refer to the earlier description in Section 3.6.3 if you need a refresher on how to do this in PDFGUI. Give your fit a descriptive name and save the project for safekeeping. We recommend frequent saves, which gives quicker recovery from accidental program crashes.

Referring to Section 3.6.4 as appropriate, load the structure from the modified CIF. Verify that the phase information has been populated correctly. Any necessary changes can be made directly within PDFGUI. Add the dataset in the same way as outlined in Section 3.6.3, and verify that it has read the correct scatterer type (x-ray) and Q_{max} value in the `Configure` pane. Change the fit range to be from 1.5 to 50 Å.

9.6.5 Set up parameters

In the data set `Configure` pane, set the appropriate values for Q_{damp} and Q_{broad} (listed in Table 9.1). Next, all of the parameters that will be refined during the modelling of the PDF as variables need to be set. As usual, we first set the phase scale factor as a variable, which can be given an initial value in the phase `Configure` pane and constrained in the phase `Constraints` tab. The structural parameters also need to be added as variables. Select the phase in the fit tree, and navigate to the phase `Constraints` pane. The lattice parameters should be set as variables, making sure, as before, to use a different number after the @ for each independent variable. Given that in this case the structure of interest is cubic, the three lattice parameters should be described by the same variable. To set the parameters for the atomic positions and ADPs, one can take advantage of a shortcut using the space group information as we did in Section 6.6.6. When in the phase `Configure` pane, select the `elem` column such that all the atoms are highlighted, then make a right mouse click on the highlighted region and select `Symmetry constraints` from the drop-down window. From the menu that appears, make sure the space group is correct ($Fd\bar{3}m$ in this case), and select the two options for constraining the positions and temperature factors. Hit `OK`, and the program will automatically generate the appropriate variables that obey the space group symmetry. Finally, note that since the local structure strongly deviates from the cubic model, the fit becomes unstable if the correlated motion parameters δ_1 or δ_2 are refined. For this reason it is recommended that one of them is selected, assigned a reasonable value, and kept fixed throughout this exercise, e.g. setting the δ_1 parameter to a value of 2.0 Å. As before, if you want to see how we set up the refinement, you can peek at our `.ddp3` file distributed along with the data, but independent work is strongly encouraged.

9.6.6 Adjust parameters and execute the fit

It is a good practice to first calculate the PDF based on the initial model and compare it to the data, as done in Section 3.6.7, in order to adjust the parameters to more realistic starting values, helping the convergence of the refinement. The first step is to calculate the PDF from the starting structure. Select your fit in the `fit tree`, then go to the `Calculations` tab at the top of the PDFGUI window and select `New calculation`.

The calculation will appear in the `fit tree`. Select the calculation, and make any necessary edits in the `Calculation Configuration` pane reflecting realistic experimental conditions. In particular, specify the correct scatterer type, adjust the values of Q_{max}, Q_{damp}, and Q_{broad} (to 27 Å^{-1}, 0.038 Å^{-1}, and 0.018 Å^{-1} respectively, in this case), and specify the calculation range (1.5 to 50 Å here). Next, press the blue gear icon in the PDFGUI window to execute the calculation. To view the calculation, select the calculation from the `fit tree`, select r as X in `Plot Control`, and then press the `Plot` button in `Plot Control`. You can also add the data to the plot by selecting the data from the fit tree and pressing the same `Plot` button. Visually compare the calculation and data to make sure they look approximately the same. In this case, we notice that the magnitude of the calculated PDF is about two times larger than that of the data, so we can set the initial scale factor to 0.5. However, by doing so one immediately realizes, upon re-calculation and comparison to the data, that the intensities of PDF peaks at small interatomic distances are not well-matched after this adjustment and that the data peaks at large interatomic distances are broader than those of the simulation peaks. This broadening arises from disorder present in the material, which should affect the observed ADPs. It is therefore recommended to increase the scale factor to an initial value of 0.8 and let the fit take care of the rest. Further, by carefully inspecting the peak positions in the simulation and in the data, we also observe that the data peaks appear at lower interatomic spacing than corresponding simulation peaks. This is not a coincidence: the ionic radius of Cr^{3+} is smaller than that of Ir^{3+}, and substitution of Cr for Ir in the material results in a lattice contraction that is not present in the starting model. Our initial lattice parameter used in simulation is obtained from the CIF file that corresponds to CuIr$_2$S$_4$ and is therefore larger than it should be for CuIr$_{1.76}$Cr$_{0.24}$S$_4$. Since this is a cubic system, it is straightforward to adjust the lattice parameter to a more realistic initial value, in this case 9.80 Å rather than 9.85 Å.

After the adjustments are done, you are now ready to refine the parameters of the model. To keep the `fit tree` tidy, you can either delete the calculation or copy the fit to a new one in which you will keep the phase and the data but not the calculation. If necessary, refer back to Section 3.6.7 for a reminder about how to run refinements in PDFGUI. As discussed in Section 6.8, the refinement process works best when a gradual approach is taken by first refining the parameters that have the biggest effect on the calculated PDF, then adding additional variables until the best possible fit is achieved. Variables can be fixed and released using the checkboxes in the `Parameters` pane of the fit, and the refined variable values can be copied to the initial parameter values for each new cycle of refinement in the way described in Section 3.6.7. However, given the high symmetry of the model in this case, there is a relatively small number of parameters, so fine guiding of the fit is probably not necessary. As mentioned previously, in this example it is not possible to obtain a good fit over the entire range given that the local and average structures are *not* consistent. The fit will result in quite large R_w of around 25%, reflecting the inability of the model to fit PDF data for short interatomic separations. However, in this specific case the misfit is highly desirable. It is precisely this failure of the cubic model that allowed us to discover the local symmetry-breaking within a material having a high-symmetry average structure.

9.6.7   Examine the wide-range fit

Once the best fit over the range from 1.5 to 50 Å has been achieved, plot the result by selecting the data set from the `fit tree` and pressing the `Plot` icon on the top task bar. Before you plot the result, make sure that in the `offset` field in `Plot Control` section you set an ample offset, e.g. −5. Once the fit result is plotted, which should look similar to that shown in Figure 9.2, examine the fit quality over different r-ranges by utilizing the `Pan` and `Zoom` tools from the plotting window. The key observation comes from the difference curve. This reveals that the cubic model provides a good fit at interatomic distances larger than ~20 Å, evidenced by a featureless difference containing exclusively low-amplitude statistical noise. On the other hand, the fit is rather poor at interatomic distances below ~10 Å, evidenced by a structured signal of substantial amplitude and indicative of a local structure with lower symmetry than that assumed by the model. Inspection of the very local features in the PDF reveals various interesting aspects, such as the existence of an Ir–Ir dimer peak at ~3 Å in the data which is absent in the model, or the shift to the left of the strong Ir–Ir peak at around ~3.5 Å in the model PDF profile as compared to the experimental data.

One possible way of rationalizing this observation is to portray it as a system featuring short-range ordered dimers spanning nanoscale domains that average out to a disordered "apparently" non-dimerized cubic structure, as observed by the average structure probes. Another possibility is that the local structure of the system is dimerized, but the dimer correlations rapidly fade out for larger interatomic distances. We are not concerned here with the correctness of these two views as the

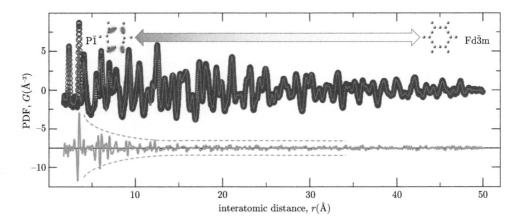

Figure 9.2 Fit of cubic $F d\bar{3}m$ model to $CuIr_{1.76}Cr_{0.24}S_4$ data at 100 K. The difference curve reveals the presence of distortions indicative of local broken symmetry due to dimerization. Short-range structure is $P\bar{1}$-like, while long-range structure is $F d\bar{3}m$.

exact interpretation is likely to be more complex, involving results of other complementary experimental and theoretical approaches. Whatever the case may be, the following statement does hold: when the experimental PDF of $CuIr_{1.76}Cr_{0.24}S_4$ at 100 K is interrogated along the r-axis, the structural information in the data will change from that corresponding to a distorted dimerized $P\bar{1}$ structure at low r values to that corresponding to an undistorted non-dimerized $Fd\bar{3}m$ structure at high values of r. It is of interest to establish a methodology for estimating the characteristic length scale of the local broken symmetry. To that effect, an r-dependent fitting protocol is utilized, in which the same structure model is implemented to fit the same experimental data set over different ranges of interatomic distance in a box-car manner, meaning that a fixed length of refinement "box" is used, but the position of the box is shifted across the pattern, sequentially refining different ranges of the PDF. The details of the box-car size and steps are determined based on the specifics of the problem at hand. In our exercise, the length scale of 1–2 nanometers is implied from the full range fit, suggesting that a starting box size of ~3–4 Å and step size of ~1 Å may offer adequate insights. Monitoring the performance of the cubic model over variable r-ranges offers an opportunity to estimate the spatial extent of local dimer correlations in the low-temperature regime of $CuIr_{1.76}Cr_{0.24}S_4$.

9.6.8   Set up and execute variable-range fits

Now that we have a good sense of how well the cubic model performs over the full data range (1.5 to 50 Å), we want to quantify how well the model performs in different r-ranges that have different information content. We will do this by slicing the full range of data into sub-ranges and running the fit in a box-car fashion. This can be done in a custom manner by copying the fit multiple times (see Section 7.6.8), adjusting the refinement range for each fit entry, and running these fits independently. It can also be done automatically within PDFGUI, the approach which we will focus on here.

The macro we will use for the r-dependent fits will use a "template" fit, i.e. a fit prepared with parameter values taken from a converged fit over the full data range. We can use the fit we prepared above for this. To start with a clean slate, it is recommended to save the prepared template as a separate project file. Before we start the r-dependent fitting, we first make sure that the template fit has the variables set up the way we want for all the r-dependent fits. Here, we will keep the same variables as in the full-range fit, except for the scale factor: it is recommended to fix the overall scale factor to the value obtained in the full data range fit as this parameter is correlated to the ADPs, which may be detrimental for the length scale estimate that we wish to carry out here. When the scale factor is fixed, select the template fit from the `fit tree`, then on the top task bar select Fits $\rightarrow$ Macros $\rightarrow$ *r*-series. The menu will now allow you to select the different fit ranges you want to include in your r series fit. To achieve this you will have to describe the desired behavior of the box car. You have to specify values for the first and last fit maximum and step (the increment), and similarly values for the first and last fit minimum and step. These should be mutually compatible and also compatible with the available data range. Here we explore a fixed fit minimum r-value and a variable fit maximum r-value protocol with 1 Å step by starting from the 1.5–5 Å range and continuing through to the 1.5–50 Å range. To achieve this, the first fit maximum is set to 5 Å, the last fit maximum is set to 50 Å, and the step is set to 1 Å. Since the template fit starts from 1.5 Å and since the fit minimum will not be changed in this protocol, the fields pertaining to the fit minimum setup of this macro in PDFGUI could be left blank. Finally, upon verifying that the desired template Fit is selected in the `fit tree`, click on `OK` on the macro pane. You will see the fit tree become automatically populated with one fit for each refinement range – 46 in total in this case. Click on one of these automatically generated fits, and you will see that these fits are linked – the initial value for each parameter is set to the refined value from the previous fit, which helps the fits start out reasonably close to the optimal structural model. To run the fits sequentially, select each of the fits (press `CTRL+SHIFT+A` to highlight all of them simultaneously) and then press the blue gear icon. It may take a few minutes for the program to run through each of the refinements sequentially, with each subsequent fit lasting a bit longer than the previous one. This is to be expected since for each subsequent fit, the PDF has to be calculated and assessed over a wider r-range than in the previous fit, taking more computational time. It is important to note that by using the r-series macro in PDFGUI, every fit in the sequence will have the same set of fixed and released parameters as they are identical to that of the template fit. If you want to make any changes to the way the fit is set up, you must go back and modify the template and generate the box-car fits again.

9.6.9   Evaluate the fit results to answer the scientific question

Once the fits are completed, we want to see how the results vary as a function of refinement range to determine whether we can establish where the local structure to global structure crossover occurs. To plot the r-dependence of a refined parameter or R_w, select one of the fits, and press `CTRL+SHIFT+A` to highlight all the others. From the `Plot Control` menu, select `index` for the X variable and whichever fitting parameter you want for the Y variable. Multiple fitting parameters can be displayed on the same plot by holding down `CTRL` and selecting the desired parameters from the Y menu. An alternative way to view the r-dependence of structural parameters is to select the structure under one of your fits in the `fit tree`, then press `CTRL+SHIFT+A` to highlight all the other structures, and then choose `index` for X and the desired structural parameter(s) for Y. Importantly, the `index` variable here represents a counter enumerating the fits in the order of appearance in the selected sequence in the `fit tree`. For example, if the selection involves all 46 files in the sequence, then `index` 0 will correspond to the first fit, `index` 1 will correspond to the second fit, and so on. This in turn means that `index` 0 corresponds to the fit maximum r_{max} of 5 Å, `index` 1 corresponds to r_{max} of 6 Å, etc. Explore the r-dependence of the fit results, and identify the parameter that will enable us to estimate the spatial extent of local dimer order in $CuIr_{1.76}Cr_{0.24}S_4$ at 100 K. Convince yourself that this is ≈ 1.5 nanometres.

9.7 Problems

1. How does the ADP of Ir evolve when increasing the fitting range maximum, r_{max}? What are the physical implications of this? Estimate the length scale of short-range Ir dimer order based on the r_{max} trend of this ADP.

2. How does the lattice parameter evolve when increasing r_{max}? What are the physical implications of this? Estimate the length scale of short-range Ir dimer order based on the r_{max} trend of the lattice parameter.

3. Compare the r_{max} dependencies of the lattice parameter and iridium ADP. Are these assessments providing a consistent estimate of the dimer order length scale, and why?

4. What is the disadvantage of using the fixed fit minimum (r_{min}) and variable fit maximum (r_{max}) protocol in estimating the length scale of local dimer order? What other protocols can be used for estimating the length scale? Compare their precision.

5. What impacts the uncertainty of a length scale estimated using r-series fitting?

6. What is the difference between protocols using a box-car of constant width and a box-car of variable width? What are the advantages and disadvantages of these different protocols?

7. How does the size of the constant-width box-car affect the observed results? What are the advantages and disadvantages of using a narrow box-car? What are the advantages and disadvantages of using a wide box-car?

9.8 Solution

In the system considered here, $CuIr_{1.76}Cr_{0.24}S_4$, the strongest x-ray scatterer is Ir, making the PDF particularly sensitive to Ir–Ir pairs. Fortuitously this is precisely the species involved in dimerization, making this example particularly suitable for illustrating the r-series fitting approach. In practice, while the states with local broken symmetry may decorate nominally high crystallographic symmetry structures of various materials, the applicability and success of the type of analysis described here will strongly depend on multiple factors, including the scattering contrast of the species involved in the short-range structure and the magnitude of the local distortions, as well as the quality of the experimental data.

The ADP of Ir increases in magnitude as the refinement range broadens (Figure 9.3, left), until it saturates in the high-r limit. This implies that the more high-r PDF data are added to the refinement, the more average-structure-like the information content becomes. The increased magnitude of the ADP at long interatomic distance results from a combination of uncorrelated atomic motion effects and enhanced disorder due to positional averaging of Ir dimers, the latter being the dominant cause. Saturation of this parameter implies that the fit is dominated by the average structure information in that interatomic distance range. The Ir ADP saturation occurs close to **index** 15, corresponding to an r_{max} value of roughly 20

Å, which can be taken as an upper bound of the short-range dimer order length scale estimated from the Ir ADP trend.

The lattice parameter displays a decreasing trend when increasing the refinement range (Figure 9.3, right), until it saturates in the high-r limit. Just as for the Ir ADP, this implies that the more high-r PDF data are added to the refinement, the more average-structure-like the information content becomes. Saturation of the lattice parameter implies that the fit is dominated by the average structure information in that interatomic distance range. The decreased lattice parameter at longer interatomic distances has complex origins which are beyond the scope of this exercise and are discussed in Božin *et al.*, (2019). They can conceptually be rationalized as follows. At shorter interatomic distances, the signal is dominated by the local environment of Ir (Cr is roughly a three-times weaker x-ray scatterer than Ir), and the response resembles $CuIr_2S_4$, which has a longer lattice parameter. At higher interatomic distances, due to directional averaging of the Ir dimers, the effect of the Cr substitution is more apparent since smaller Cr defects introduce a lattice contraction. The value of the lattice parameter saturates close to **index** 25, corresponding to an r_{max} value of roughly 30 Å, which can be taken as an

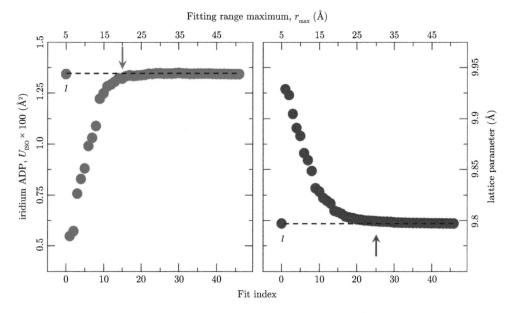

Figure 9.3 Evolution of Ir ADP (left panel, left ordinate) and lattice parameter (right panel, right ordinate) values with maximum fitting range, r_{max}, obtained from $Fd\bar{3}m$ fits to $CuIr_{1.76}Cr_{0.24}S_4$ data at 100 K. The x-axis represents the fit index, as discussed in the text. The isolated points represent values obtained from wide-range fits. Horizontal dashed lines are guides to the eye. The vertical arrows mark estimates where the considered parameters deviate visibly from their saturation values.

upper bound of the short-range dimer order length scale estimated from the lattice parameter trend.

It is clear that the trends of the two parameters are similar in that they both saturate at high interatomic distances. However, the saturation is reached at different **index** values, and in turn different $r_{\max}$ values. This is presumably due to different physical origins of the saturation of these two parameters. The ADP saturation originates from PDF peak broadening, which is driven by the dimer disordering affecting PDF peaks at high r in a similar way. The lattice parameter saturation originates from PDF peak shifts, which are caused by lattice contraction induced by random Cr defects. The shifts gradually increase with r. The influence of the PDF peak position weight in the fitted data therefore lags behind the influence of the PDF peak width weight, causing this observed behaviour. The ADPs provide a more conservative estimate as they more directly couple to the dimer disordering.

The approach for estimating the lengthscale of local order based on a fixed $r_{\min}$ and variable $r_{\max}$ is a variable-width box-car protocol. In this case, the information density assessed by the model is not uniform for all fitting ranges. Furthermore, since $r_{\min}$ does not change, the portion of data pertaining to short-range order will participate in fits for every range, which may bias the structure analysis, particularly if quantities such as R_w are considered as a function of the fitting range. An alternative protocol that can be used is one where $r_{\max}$ of the fit is kept fixed, while $r_{\min}$ varies. This is also a variable-width box-car protocol, but in this case the short-range symmetry broken heterogeneity is approached "from above". This latter method arguably provides better sensitivity of R_w to the local deviations from the average structure as the refined structural parameters will still be dominated by the average structure, while R_w will exhibit an upward deviation from its horizontal trend with decreasing $r_{\min}$ as soon as pieces of PDF data containing information on local deviations from the average model become part of the fit. An advantage of a variable $r_{\min}$ and fixed $r_{\max}$ protocol is that the data density is relatively uniform. In the low-r region of the PDF, the peaks are fairly well separated, and the information density is low. At higher r values, on the other hand, PDF peaks overlap significantly, giving a much higher information density. This high information density at high r values also allows for finer box-car steps when having a variable $r_{\min}$ rather than $r_{\max}$.

One can also use a protocol in which both $r_{\min}$ and $r_{\max}$ are varied, typically with a fixed box-car width. In the variable-sized box-car approaches, larger box-car sizes will probe both short-range and long-range structural information from PDF data, a coupling that could hamper the length scale estimate, and the box-car width should therefore be smaller than the length scale of interest. In this case, a fixed-width box-car approach can be advantageous as it decouples portions of PDF data with conflicting information content and can have higher sensitivity to structural changes. However, if the width of such a box-car is too small, then the extracted parameters will probably be subject to fluctuations originating from variable information density in different PDF ranges. This will be particularly pronounced when such a narrow box-car rasters the low-r region of the PDF where the peaks are well separated. Furthermore, narrow box-car fits will also be more vulnerable to correlations of refined parameters. Wider fixed-width and variable-width box-car protocols are generally less prone to such artefacts, and the length scale can be more cleanly defined, albeit at the expense of higher uncertainty in the extracted correlation length and lower sensitivity to the r-dependent changes in the signal.

It is important to keep in mind that the analysis described here offers a tool for *estimating* characteristic length scales, which often involves a crossover from a short-range ordered local broken symmetry to a global higher symmetry structure. It is therefore recommended to employ multiple approaches involving both variable- and constant-width box-car fitting protocols, as well as different box-car steps. Such systematics can also provide some insight into the uncertainty of the length scale estimate. It is worth noting that if, instead of a continuous crossover, one deals with coherent domains, then the short r-range PDF will be dominated by intra-domain structural correlations, whereas the long r-range PDF will be dominated by inter-domain structural information. In such a case, the domain boundary may be sharply defined, but the non-spherical domain shape and presence of finite-sized domain walls of possibly different structures will hamper the analysis, and more complex approaches will be needed. Similar difficulties apply in cases of strongly anisotropic crossovers. However, the general analysis principles involving r-dependent fitting presented here may still provide useful insights.

9.9   Diffpy-CMI solution

This section describes the Diffpy-CMI solution. It may be skipped on a first reading, but it will be helpful to you when, in the future, you want to learn how to use Diffpy-CMI. To get started with Diffpy-CMI, follow the instructions in Section 3.9. The Diffpy-CMI code for the solution will be in a Python file in the diffpy-cmi folder under solutions for this chapter in your downloaded data. Read through the comments there. They are expanded on below in greater detail where needed.

9.9.1   Running Diffpy-CMI examples

See the instructions in Chapter 3 for help on running Python files.

9.9.2   `fitRSeries.py`

This file presents a Diffpy-CMI solution to the refinement of a series of r-ranges with PDF data collected from $CuIr_{1.76}Cr_{0.24}S_4$ at 100 K. Our goal is to estimate the spatial extent of local dimer order, just as you did using PDFgui above.

To run this code, you should have downloaded the script in a directory called Diffpy-CMI, with a parent directory called `solutions` and another directory called `data` located at the same directory level as `solutions`. The `data` directory should contain the files `Fd-3m.cif` and `CuIr1.76Cr0.24S4-q27r70t100-11IDC-APS.gr`. This example will again take some time to run, as we are running many PDF fits. Have a look at `fitRSeries.py` and the comments therein. We will only explain here the numbered comments representing significant deviations from the previous DIFFPY-CMI examples.

 # 1-5: We again begin by importing the necessary packages and giving some details about the measurement, as well as details on the r-range we'd like to fit over when we consider the entire PDF.

 # 6: In this example we will consider many different r-ranges in three situations: fitting the full PDF, fitting with a sliding box-car window, and fitting with a sliding r_{max}. Here we identify the total number of r-ranges for the last two cases, as well as the window size for our sliding box-car fits.

 # 7: Using the information from the previous item, we construct an array of numbers representing the lower bounds, r_{min}, for our sliding box-car fits.

8: We then take this array and add to each entry our box-car window size to get an array representing the upper bounds, r_{max}, for our sliding box-car fits.

 # 9: These two new arrays get combined into a list of tuples, each of the form (r_{min}, r_{max}). The list contains one tuple (tuples are Python data structures that resemble lists but are immutable) for each r-range of our sliding box-car fit, for convenient use later on.

 # 10: We can reuse the r_{max} array to build a similar list of tuples for our sliding r_{max} fits, where the lower bound r_{min} remains fixed and we vary the upper bound only.

 # 11: As we mentioned previously, we want to handle three fitting situations: fitting the full PDF, fitting with a sliding box-car window, and fitting with a sliding r_{max}. Here, we build a list, where each entry is a list of tuples we created in the previous steps.

 # 12: We would also like to have some awareness within our code of just which of the three fitting situations we are currently working with, so we make a list of strings, where each entry is a description of the fitting situation.

 # 13-33: From here, our example proceeds much as in previous cases. We set some initial values for our parameters, notably the occupancy of the Ir site as the structure also contains Cr on this site. We define our `make_recipe` function, and as in the DIFFPY-CMI example in Chapter 3, we need to add atoms to the loaded structure and modify their occupancy, define our own isotropic ADP parameters, and use them to constrain our fit. Just as in the PDFGUI example in this chapter, both Ir and Cr together share a single ADP fit parameter. We also create our `main` function, which begins with a call to `make_recipe`.

 # 34: In this script, we will handle all three fitting situations by using Python to loop over the different lists of (r_{min}, r_{max}) tuples we built earlier, rather than hard coding each situation. Each list of tuples is paired with the string description of the fitting situation we defined in previous steps. We have already called `make_recipe` prior to entering any loop, and we will not call it again within the loops, so each iteration will begin with the fitted parameters of the previous loop. The first iteration of the loop will be the full r-range, such that we have a good starting point for the refined parameters when we begin considering narrower r-ranges.

 # 35: We loop over each (r_{min}, r_{max}) tuple in the list so that we consider all possible r-ranges for each situation.

 # 36: We use the upper and lower fitting bounds on r to build a name for our fit.

 # 37: At each step of our inner loop, we need to be sure to change the r-range we are fitting over.

 # 38–39: Our three fitting situations can be further subdivided into two categories, either considering the majority of the available r-range or considering a subset of this r-range. As in the PDFGUI example of this chapter, we will refine a complete parameter set in the former case and a partial subset of parameters in the latter case. We can handle this using an `if...else` statement when defining our list of parameters and/or tags to free sequentially.

 # 40–44: The remainder of our code proceeds identically to the previous examples. We optimize our fit recipe by freeing parameters sequentially, and then we write results using our bespoke fit name. Each time the loop on our r-range rolls over, the recipe will retain the best-fit values from the previous iteration of the loop.

Once this has finished, your directory will be populated with plots of the best fit for each r-range, as well as text files describing the fitted parameters and fit statistics. These files can be post-processed and parsed using Python to generate plots of parameters as a function of r-range, or this could be done right inside our fitting script with a few modifications.

10

Nano- and polycrystalline thin films: Local structure of nanocrystalline TiO$_2$ grown on glass

Maxwell W. Terban

10.1 Introduction and overview

In the previous chapters, we have mostly dealt with PDFs from strongly scattering samples, such as bulk crystalline materials or nanoparticles made of high-Z elements. We have also only looked at data collected under ideal conditions, i.e. where a powder of the material has been packed into a capillary made of a weakly scattering amorphous material, which contributes very little to the measured signal. However, many experiments may not provide for such ideal samples and conditions. For example, we may want to collect data from nanoparticles or clusters in solution, where the scattering signal from the solvent is much larger than that of the particles themselves. We could also collect data from a catalyst on a support material, where the signal from the catalysts is again much smaller than that of the support. Sometimes, we must also compromise the x-ray exposure time, and therefore the data quality, for the sake of fast time resolution during *in situ* or *operando* experiments. In this chapter, we will show how we can treat data where the background contribution to the scattering pattern is significant and where the data are noisier than we have previously seen.

This chapter presents an example of using the thin film PDF (tfPDF) technique (Jensen *et al.*, 2015). Thin films are a class of materials where the limited out-of-plane versus in-plane dimensions can result in special properties and applications. Structure characterization of thin films can be difficult due to the small amount of sample present on the film substrate, and grazing incidence x-ray diffraction methods are usually applied for characterization of crystalline films as this method maximizes the film-to-substrate signal. It is possible to perform PDF analysis using rather specialized grazing incidence x-ray total scattering measurements (Dippel *et al.*, 2019), but the measurements are quite complex and require a specialized setup at the synchrotron beamline. However, the high flux of synchrotron x-ray

Atomic Pair Distribution Function Analysis. Maxwell W. Terban, Oxford University Press.
© Simon J. L. Billinge and Kirsten M. Ø. Jensen (2023). DOI: 10.1093/oso/9780198885801.003.0010

beams combined with recent developments in data reduction software imbues very high signal sensitivity in PDF measurements, allowing large background scattering contributions to be subtracted while maintaining high quality data (Terban *et al.*, 2015). This allows for x-ray total scattering data from the films to be measured directly in normal incidence transmission, as is the case for standard capillary measurements. Here, we show how to subtract large backgrounds from the raw data; we then demonstrate how the thin film structures can be subsequently characterized using similar approaches to the materials discussed in previous chapters.

10.2 The question

TiO$_2$ thin films are important in a wide range of applications, including functional coatings, solar energy production, and photocatalysis (Diebold, 2003). The processing conditions can enable high-level synthetic control over the resulting structure and properties of the films. In this example, we are concerned with the product obtained from microwave-assisted TiO$_2$ thin film synthesis. Glass substrates coated with indium tin oxide (ITO) are immersed in a growth solution containing dissolved precursors and then irradiated with microwaves (Reeja-Jayan *et al.*, 2012). Under microwave irradiation, TiO$_2$ growth happens at significantly lower temperatures compared to conventional growth methods, presumably by the strong absorption of microwaves by the ITO layer, though the exact mechanism is not fully understood (Nakamura *et al.*, 2021).

Our primary goal is to use the PDF to gain insight into the structure of the TiO$_2$ films. Three common polymorphs of TiO$_2$ exist: rutile, anatase, and brookite. They differ in their atomic arrangement by the patterns of corner and edge sharing of [TiO$_6$] octahedra, as shown in Figure 10.1.

The anatase form is typically preferred for most applications, due to its higher photocatalytic activity (Sclafani and Herrmann, 1996). The nanocrystalline nature

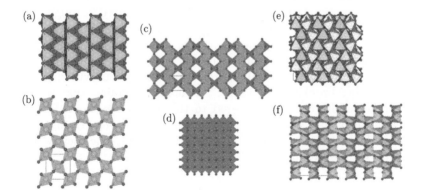

Figure 10.1 Structure of (a) rutile seen along the [100] direction and (b) [001] direction, (c) anatase seen along the [010] direction and (d) [001] direction, and (d) brookite seen along the [100] direction and (e) [001] direction. Titanium and oxygen atoms are represented by spheres/polyhedra and red spheres, respectively.

of the films can make it difficult to identify which structures have formed. Therefore, PDF analysis can be helpful to determine under which conditions anatase forms, when and what other products form, or even how the composition and nanostructure vary spatially across the film.

We note that this example of thin-film PDF (tfPDF) works because the nanocrystalline films do not have any crystalline texture or preferred orientation. Recently, approaches have been derived and demonstrated that allow samples with a preferred orientation of the grains to be studied using PDF methods (Gong and Billinge, 2018; Harouna-Mayer *et al.*, 2022; Cervellino and Frison, 2020a).

10.3 The result

The glass substrate accounts for about 99% of the total scattering signal. Despite this, a good quality structure function and PDF can be successfully extracted by careful subtraction of the background signal before the Fourier transform. It is shown that the same result can be obtained by subtraction of the separately processed PDFs in real-space since both representations contain the same information. However, this method can be less straightforward and results in slight differences in the treatment of systematic errors.

PDFs are used to compare the structures of films synthesized from microwave-assisted and conventional growth methods. These include a microwave-assisted film grown at 160 °C and 40 W microwave power for 60 min, and two furnace-grown films, which were heated up to either 250 °C or 450 °C over one hour, then held at temperature for an additional hour (Nakamura *et al.*, 2017).

The structure of the microwave-assisted TiO_2 film is nanocrystalline anatase with a structural coherence length of approximately 49 Å. It was found that the residual from a single-phase fit contains a significant structural signal. This residual can then be described by an amorphous component that shows similarity in structure to brookite but with a coherence length of only about 11 Å.

Through the same steps, PDFs for the films produced through conventional growth methods were extracted. Varying structures were found to form in the conventionally grown films depending on the processing temperature. With the conventional synthesis method, processing at 250 °C does not give a crystalline TiO_2 product, the substance remaining relatively amorphous with a structure most closely resembling brookite. At 450 °C, a more crystalline anatase structure is formed. These results can be compared to the microwave-assisted case in which anatase was achieved at the much lower temperature of 160 °C, and this temperature can be lowered even further to increase the energy efficiency with which the anatase films can be produced.

10.4 The experiment

The geometry of the measurement is similar to standard transmission experiments. However, instead of a capillary, the whole film plus substrate is used. The films are mounted perpendicular to the incident beam using a bracket for flat plate samples, using tape to hold the film and substrate in place. The whole bracket plus sample is mounted on the goniometer so that the beam passes through the substrate before hitting the thin film. This setup is shown in Figure 10.2 and described in Jensen *et al.* (2015) and Nakamura *et al.* (2017). See Table 10.1 for the experimental information and Table 10.2 for the files to download.

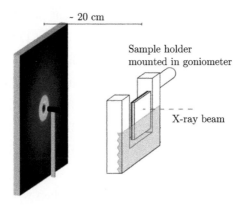

Figure 10.2 Schematic drawing of the setup used for tfPDF. Reproduced with permission from Jensen *et al.* (2015).

Table 10.1 Experimental conditions.

Facility	NSLS-II
Beamline	XPD
Detector type	Perkin Elmer amorphous silicon 2D detector
Sample geometry	Film on 0.5 mm thick substrate of glass plus ITO
Sample environment	Room temperature, ambient conditions
Experiment	**Microwave films/Conventional films**
x-ray wavelength	0.1827 / 0.1835 Å
Exposure time	1800 / 900 s
Q_{damp}	0.0437 Å^{-1}
Q_{broad}	0.0170 Å^{-1}

10.5 What next?

1. Download the files. Plot the diffraction patterns and target PDFs to see what they look like.
2. Use PDFGETX3 or XPDFSUITE to generate PDFs.
   (a) Load the `ito-glass.chi` and subtract `glass.chi` as a background to extract the $I(Q)$, $F(Q)$, and $G(r)$ of the ITO layer.

Table 10.2 Files to download.

Filename	Note
Microwave grown film	
`tio2-ito-glass.chi`	measurement of TiO_2 on ITO-glass
`ito-glass.chi`	measurement of ITO-glass substrate
`glass.chi`	measurement of glass subtrate
`template.cfg`	A template config file for PDFGETX3
`ito-minus-glass.gr`	Target PDF for ITO layer
`tio2-ito-minus-glass.gr`	Target PDF for ITO/TiO_2 layers
`tio2-minus-ito-glass.gr`	Target PDF for TiO_2 layer
Extra processed files	
`tio2-ito-glass_noBKGsub.gr`	TiO_2-ITO-glass w/o background subtraction
`ito-glass_noBKGsub.gr`	ITO-glass w/o background subtraction
`glass_noBKGsub.gr`	glass w/o background subtraction
`tio2-ito_minus-glass_TiO2Comp.gr`	ITO/TiO_2 layers w/ TiO_2 composition
`ito-glass_TiO2Comp.gr`	ITO-glass w/ TiO_2 composition
Conventionally grown films	
`tio2-ito-glass-250.chi`	measurement of TiO_2 (250 °C) on ITO-glass
`tio2-ito-glass-450.chi`	measurement of TiO_2 (450 °C) on ITO-glass
`ito-glass-conv.chi`	measurement of ITO-glass
`tio2-ito-glass-250.gr`	Target PDF for 250 °C conventional TiO_2 layer
`tio2-ito-glass-450.gr`	Target PDF for 450 °C conventional TiO_2 layer
Structures	
`TiO2-anatase.cif`	CIF format structure of anatase form TiO_2
`TiO2-brookite.cif`	CIF format structure of brookite form TiO_2
`TiO2-rutile.cif`	CIF format structure of rutile form TiO_2
`ITO-bixbyite.cif`	CIF format structure of bixbyite form ITO
`ITO-PR7O12.cif`	CIF .cif format structure of Pr_7O_{12} form ITO

Filename	Note
Code	
`scaling_code.py`	Code to autoscale PDF contributions
`background-scale_in_real-space.py`	Determine background scale in real-space.
`fit_tio2_ito_contributions.py`	Scale TiO$_2$ and ITO to the glass-only subtracted PDF

    (b) Adjust the scaling of the background; observe when the Bragg peaks from ITO come into view in $F(Q)$ as you increase the background scale starting from zero.

    (c) Pay careful attention to how $G(r)$ changes as the background scale is increased.

    (d) Determine the optimal background scale, and optimize the Q_{max} to reduce the amount of noisy signal considered.

    (e) Save the data as a `.gr` file.

    (f) Repeat the last five steps to obtain the PDF for the ITO and TiO$_2$ layers together:

        `tio2-ito-glass.chi` with `glass.chi` as background.

    (g) Repeat for the microwave-assisted TiO$_2$ layer separately:

        `tio2-ito-glass.chi` with `ito-glass.chi` as background.

    (h) Repeat for the TiO$_2$ layer from conventional processing at 250 °C:

        `tio2-ito-glass-250.chi` with `ito-glass-conv.chi` as background.

    (i) Repeat for the TiO$_2$ layer from conventional processing at 450 °C

        `tio2-ito-glass-450.chi` with `ito-glass-conv.chi` as background.

    (j) Exit PDFGETX3/XPDFSUITE.

3. Instead of subtracting the background signal from $I(Q)$, process the PDFs for `tio2-ito-glass.chi` and `ito-glass.chi` with no background subtraction and the chemical composition set to TiO$_2$ for both. Subtract the resulting PDFs to isolate the TiO$_2$ signal, and compare to the previous result.

4. Get a CIF file from PDFITC to use as a starting point

5. Compare the PDF of TiO$_2$ + ITO to the separate TiO$_2$ and ITO PDFs.

6. Using anatase, brookite, and rutile structure models, use PDFGUI to characterize the structure of the microwave-deposited TiO$_2$ film:

    (a) Try single-phase fits.

    (b) Try multi-phase fits.

7. Characterize the structure of the conventionally grown TiO$_2$ film processed at 250 °C.

8. Characterize the structure of the conventionally grown TiO$_2$ film processed at 450 °C.

9. Use PDFGUI to compare the PDF obtained from the ITO layer to the different ITO structures provided.

10.6 Wait, what? How do I do that?

10.6.1 Download the files and plot the data files and target PDF files to see what they look like

Follow the instructions in Section 3.6.1.

Start by plotting the `tio2-ito-glass.chi`, `ito-glass.chi`, and `glass.chi` data on top of each other. Plot the files using your preferred plotting program. See Section 4.4 for help with `.chi`, `.iq`, etc. file formats. See Section 4.6.2 for help with doing this using Diffpy tools. You will see that on a first inspection, the data look very similar as the scattering intensities in all three are completely dominated by the contribution from the glass.

You can also plot the data from the samples prepared by conventional synthesis methods, and it is also a good idea to plot the target PDFs provided.

10.6.2 Start PDFGETX3 or XPDFSUITE and load the data files

We described the basic steps for using PDFGETX3 and XPDFSUITE in Chapter 4, so refer back to that if you need a reminder. We will start by treating the `ito-glass.chi` files with `glass.chi` as background.

As written in the file headers, the scattering intensities are given as a function of Q in nm^{-1}. The composition of ITO is $In_4Sn_3O_{12}$. The data are collected using a square 2D detector, so we should not take the intensity at the highest Q values into account (as these are measured only in the corners of the detector). As we will see when we have subtracted the background, the signal from the sample is rather noisy, and in this case, it can be a good idea to set $Q_{\text{max-inst.}}$ to a lower value, e.g. 20 Å^{-1}. You can keep the value of *rpoly* at 0.9 Å. Q_{min} can be set to 0.7 Å^{-1}, and for now, set the value of Q_{max} to 20 Å^{-1}. We will adjust Q_{max} further below.

10.6.3 Find the correct background scale

You already saw in Chapter 4, that due to differences in measurement time and slight fluctuations in beam intensity during the synchrotron experiments, we often need to adjust the background scale in order to find the optimal subtraction of the background scattering signal.

Since the intensities measured from the substrate here account for about 99% of the total scattering signal, we essentially only "see" the structure of glass if the background scale is set to zero. Try this out for the `ito-glass.chi` data with the `glass.chi` background by using the slide bars to set the background scale to zero. You will see a PDF showing only short-range order, as expected from amorphous glass. At this point, it is a good idea to note where the most intense glass peaks are in the $F(Q)$ and $G(r)$ functions. For example, glass has a very intense PDF peak at ~ 1.6 Å, arising from Si–O distances. When we subtract the background from our sample data, we basically want to minimize this peak as we do not expect ITO or TiO_2 to show PDF peaks in this region.

As the total x-ray exposure time used during the data collection was the same for the `ito-glass.chi` and `glass.chi` data, a good starting value for the background scale is 1. With this value, we start seeing the signal from ITO in the $F(Q)$ and $G(r)$ functions. However, there is still a signal from glass, for example, the very intense Si–O peak is still clear in the PDF. If you zoom in on the $I(Q)$ data, you will see that the data for the sample and background do not fully line up. We basically want the lines to overlap except for at the values in reciprocal-space where ITO shows Bragg peaks, which you will see as minuscule differences when you zoom in. Figure 10.3 shows this for the `tio2-ito-glass.chi` data.

Try to adjust the background scale value to remove the contributions from glass. You will see that the $F(Q)$ and $G(r)$ functions are very sensitive. We found that a value of ~ 1.013 worked well here. When finding a good value for the background scale, it is important to check $F(Q)$ and $G(r)$ as it is here where the amplified signal makes it clear whether the glass signal has actually disappeared when the background scale is sufficiently close to the correct value.

In this thin film case, the background signal is much stronger than the signal of interest, and the background subtraction is very challenging. PDFGETX3 is well suited to this situation because the ad hoc correction allows very precise background subtractions even if the measurement of the background is not perfectly reproduced. However, the ad hoc correction can result in some interesting and non-intuitive behaviour of the functions. We discuss some of these "gotchas" briefly here (gotchas are slightly unexpected software behaviours that can trip up a new practitioner).

First, we mention that close to the ideal subtraction point, where the main signal and the background are almost identical, very small changes in the background scale can have very large effects on the resulting $F(Q)$ and $G(r)$. As you close in on the right solution, you will find that you need to vary the background scale to more and

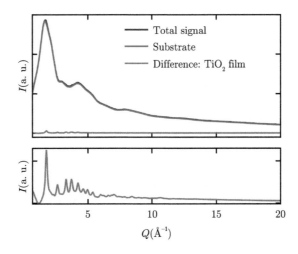

Figure 10.3 The total sample $I(Q)$ shown overlaid with the scaled $I(Q)$ from the substrate with difference below.

more significant figures. For example, in the case here, we found that varying the third significant figure of the background scale gave a meaningful improvement in the extracted $G(r)$. If you are using XPDFSUITE, there is a zoom feature associated with the background scale slider (it looks like two magnifying glasses), which helps with this process. Initially you can have a coarse slider, allowing the full-scale of the slider to vary by one significant figure. As you close in on the right scale, you can zoom in so that the full-scale of the slider is in the second significant figure, then the third, and so on. If you get in too tight too quickly, you can zoom out and then zoom back in, and so on. This takes some getting used to, but it is a feature that is very helpful and can be a practical addition to your workflow.

Second, we note that close to the point where the signal and the background are very similar, the scaling of the $F(Q)$ and $G(r)$ become very non-linear. You may find that the scale factor of the extracted $F(Q)$ starts to grow in scale very rapidly and may even invert with a large negative scale. This behaviour is expected because of the nature of the ad hoc correction and does not indicate anything wrong with the correction. The thing to remember is that, by design, PDFGETX3 returns $F(Q)$ (and $G(r)$) accurately, except for an arbitrary scale factor (Juhás *et al.*, 2013). We find that this scale factor is often close to 1 when the background signal is much smaller than the signal from the sample. However, when the background and sample signals are very similar and the signal of interest is a small percentage of the total signal, this arbitrary scale can vary widely. It does not mean that there is anything wrong with the extracted $F(Q)$. If you encounter this, do not worry. Look for some feature in $F(Q)$ or $G(r)$ that is an indicator of your background (it might be a peak in $G(r)$ at ~ 1.6 Å coming from the silica substrate, for example, as in the current case) and vary the background scale until this feature disappears in $G(r)$, ignoring what is happening to the scale of $G(r)$.

Finally, we note that your best background subtraction may actually occur when the signal has inverted and the peaks from your desired signal are upside-down. This seems to be counterintuitive and a sure sign that you have over-subtracted the background. However, this is not necessarily the case. As mentioned above, PDFGETX3 only promises to give you the best PDF it can to an arbitrary value of the global scale factor. Since the value of the scale factor is arbitrary, it can also take negative values, which will invert the PDF peaks. To revert it to being upright, simply multiply $F(Q)$ and $G(r)$ by a negative number. In general, an inverted $G(r)$ might indicate that you have over-subtracted your background; in cases when a small signal is being extracted from a very large background (thin films or clusters in solution, for example), the best background subtraction can sometimes happen in this inverted regime. The physical origin is that the measured background does not perfectly reproduce the background signal in the sample. In this case, the ad hoc correction can perhaps find a better solution that perfectly removes a structured signal (such as the silica substrate) when the unstructured part of the background (for example, from air scattering) in the measured background exceeds that in the sample scattering signal.

In the latest version of PDFGETX3, it is possible to subtract multiple backgrounds, which can help you do a better job in this situation. To more accurately account for the different components of the background, you could measure just the air scattering, or an empty sample chamber, then make a separate measurement of an empty sample holder/substrate, and finally measure the sample. It may then be possible to get an even better background subtraction by removing a linear combination of the empty container and the air scattering from the sample.

10.6.4 Optimizing the Q_{max} value

With the background subtraction completed, you will see that the PDF is quite noisy. This is most clear when considering the $F(Q)$ function, where lots of noise is present at high Q values. We therefore want to adjust the Q_{max} value to optimize the PDF. Try to use the slide bars and see the effect of adjusting the Q_{max} value. A value of ~ 16 Å^{-1} is a reasonable compromise between noise and resolution for all datasets here. You will find a noisier signal for the ITO layer because it is thinner than the TiO$_2$ layer on the film.

Now you can repeat the process for the other data files. You want to obtain PDFs using:

1. `ito-glass.chi` with `glass.chi` as the background: ITO layer.
2. `tio2-ito-glass.chi` with `glass.chi` as the background: ITO and TiO$_2$ layers together.
3. `tio2-ito-glass.chi` with `ito-glass.chi` as the background: TiO$_2$ layer.
4. `tio2-ito-glass-250.chi` with `ito-glass-conv.chi` as the background: TiO$_2$ layer from conventional processing at 250 °C.
5. `tio2-ito-glass-450.chi` with `ito-glass-conv.chi` as the background: TiO$_2$ layer from conventional processing at 450 °C.

You need to adjust the background scale for each dataset individually to get a good PDF, but remember to keep your input values for *rpoly*, $Q_{\text{max-inst}}$, Q_{min}, and Q_{max} the same for all datasets as we want to compare the PDFs directly in the next steps. You can see which values we used in the headers of the `.gr` target files that you downloaded along with the data.

10.6.5 Comparing background subtraction in reciprocal- versus real-space

Before we move on to analyse the PDFs you have just obtained, we will show how the background subtraction we have just done could also be done in r-space, i.e. after the Fourier transform. In the data processing done so far, we have subtracted the background in reciprocal-space, i.e. before the Fourier transform. We want to see how the result compares to the alternate case, where we instead do the subtraction *after* the PDFs have been generated. In order to do this, we must first propagate the measured scattering intensities for the sample and background measurements through to the PDFs without doing any background correction, by setting the background scale to zero.

You can again do this in PDFGETX3/xPDFSUITE. Process the tio2-ito-glass.chi and ito-glass.chi data with the background scale set to zero. We will use the PDF from ito-glass.chi as background in the r-space subtractions, and even though the composition for this measurement is different from the sample measurement (i.e. it has no TiO_2), we should process the background PDF with the same final composition as the sample data – we simply use TiO_2 for both. The reason is that when we generate the $F(Q)$ function from $I(Q)$, we are correcting for the scattering cross section of the component elements, so we need to use a consistent composition to avoid systematic differences in the resulting PDFs. It is equally important that the other processing parameters are consistent: $Q_{max\text{-inst.}}$, *rpoly*, Q_{min}, and Q_{max}.

Once the PDFs are processed, you can plot and examine them. You should now try subtracting the PDF obtained from ito-glass.chi from the one obtained from tio2-ito-glass.chi as-is. You can do this however you want, e.g. with a Python script or by using a plotting and data analysis program of your choice. This direct subtraction (with an r-space background scale of 1) may not give a very good result, since we still need to optimize the scaling between the datasets. You can use your own method of choice to optimise the r-space background scale. You can do it manually as we did for the reciprocal-space background subtraction by simply adjusting the scaling you use in the subtraction. For example, adjust the scale while keeping an eye on the PDF peak at 1.6 Å, and check how it behaves when increasing or decreasing the scale. Since we know this peak arises from the glass background, you can find a background scale where this peak basically disappears.

You can also use the included Python code `background-scale_in_real-space.py` that you downloaded with the data files. You can find more information about running Python scripts in Appendix A. In short, the Python scripts can be run by first opening a command terminal in which you can interact with your Python distribution and navigating to the directory containing the Python scripts. Then, type

```
> python background-scale_in_real-space.py
```

and press enter. The function `scaledPDFs` will optimize the scale factors of input datasets to best fit a target dataset. The results will be plotted automatically, showing the scaled PDFs, the subtracted signal compared to the reciprocal-space subtracted dataset, and the difference between the two. The terminal output will be a two-column array with the scale factors for the two datasets, and then the residual R_w value between the fitted and target data.

10.6.6 Comparing separated PDF contributions

We can now start comparing the TiO_2 PDFs that we obtained using background subtraction in reciprocal-space versus r-space. Plot the PDFs on top of each other. What do you see? If the background subtraction has been done correctly in Q and r space, respectively, you should see quite similar results. This shows us that background subtraction can be done either before or after the Fourier transformation and still give reasonable results. We discuss possible differences between the two approaches to background subtraction and what we prefer below in Section 10.8.

As a sanity check, we can also compare how subtracting out certain contributions to the total scattering signal affects the resulting PDF data. Essentially, how does the manipulation of information content in reciprocal-space affect the information content in real-space? We have already seen that whether we subtract the substrate scattering in reciprocal- or real-space does not affect the result as long as we can determine the correct background scale. In light of this, we can also test the effects of subtracting out different components from the same dataset. This will hopefully give an idea as to the versatility of this difference PDF method, especially for multilayer samples. For example, you can test what happens if you subtract the PDF of ITO from the PDF containing both the ITO and TiO$_2$ signals. You can use the same scale optimization approach to compare the separate PDFs to the total. You can try preparing your own script, or use the provided code: `fit_tio2_ito_contributions.py`. It is once again important that the same composition (set to TiO$_2$, e.g. see additional processed files with `_TiO2Comp.gr`) is used when generating the PDFs for real-space background subtraction to avoid systematic differences. How do the results compare?

10.6.7 Get a CIF file from PDFITC to use as a starting point

Using STRUCTUREMINING in PDFITC is a great way to get CIF files to make a start on your refinement campaign. Please see 6.6.2 for more details.

10.6.8 Start a PDFGUI project and start a new fit

Now that we have extracted the target PDFs, we want to investigate the structure of the TiO$_2$ layers in the thin films. You can now choose whether you want to use TiO$_2$ PDFs obtained from reciprocal-space or r-space background subtraction – either way will work.

We will again use PDFGUI, so open it up and set up a new project, following the instructions from the previous chapters.

10.6.9 Refining TiO$_2$ polymorph structures

10.6.9.1 Single-phase refinement

We start by analysing the PDF obtained from the microwave-processed sample and focus on the TiO$_2$ layer. The first thing we want to do is to "fingerprint" the sample structure phase or phases. For crystalline materials, it is best to do this by indexing the diffraction patterns in reciprocal space, because the Bragg reflections from different crystal phases are usually well separated in Q, as opposed to real-space where all signals overlap in r. However, phases can also be identified using the PDF, with the additional benefit that the local structure in highly nanocrystalline and even amorphous phases can be compared to crystal structure models. Here, we proceed by comparing the experimental PDFs to PDFs calculated from the known TiO$_2$ structures discussed above. Try to get CIFs using PDFITC by uploading the

TiO_2 PDF to the STRUCTUREMINING app. For convenience, we have provided CIFs for these structures (anatase, rutile, and brookite), and we start by calculating PDFs from them. Import the `.cif` files in PDFGUI, and set up PDF calculations for them, as described in Chapter 3. When configuring your calculations, use the appropriate Q_{damp}, Q_{broad}, and Q_{max} values from above, and make sure that the ADPs for the structures are reasonable and at the same value for all the structures, which makes it possible to compare them directly. Now import the `.gr` file for the TiO_2 layer into your fit tree and make a plot of it next to your calculated PDFs. Which calculated PDF do the data resemble the most?

We can now fit the data using the best candidate structure. Under "Data Set Configuration", remember to input the given experimental resolution parameters, Q_{damp} and Q_{broad}, and set your refinement range to something like 1.0–30.0 Å. You can first calculate the PDF of the phase for initial comparison without any variables set by clicking on the gear in the top left of the PDFGUI window. Based on this and the symmetry of the phase, you can determine which structural parameters to constrain and carry out a basic structural refinement of the phase. If you are not already comfortable with basic structural refinements, please refer to Chapter 6. As always, think about the sequence of letting refinement variables free. Also be careful not to "overfit" using too many variables, given that the data are not of the same quality as those you have worked with in the previous chapters. As mentioned before, it can be helpful to check whether refined variables are statistically correlated by looking at the program output in the `Results` tab of the fit. In the current case, we determined that it is not advisable to fit anisotropic ADPs. You can see our results from a fit with the anatase structure if you sneak a peek at Figure 10.5 in the Solutions section. While the anatase structure gives a reasonable description of most of the PDF signal, we see a significant residual in the low-r region of the PDF that contains unfit structural information: it appears as if the data actually arise from two different phases. If you compare the residual with the PDFs calculated from the different TiO_2 phases, which PDF does it resemble? Hint: see Figure 10.5 where we show that the residual itself can actually be fitted fairly well with the brookite structure.

10.6.9.2 Two-phase refinement

We next want to see if we can improve upon the results of the previous section by trying two-phase fits of anatase and brookite to the measured data. Rather than starting from scratch, the best way to initiate a two-phase fit is to copy your single-phase anatase fit and paste below. Then input the second phase, brookite, under the same fit tree. You can refer to Chapter 5 for setting up two-phase fits.

We want to characterize the phase fraction of the two structures. As mentioned in Chapter 5, this can be done using either two independent structure scale factors, or one global scale factor and one independent phase scale factor. Here, we choose the latter option. We can first set a global scale factor under "Data Set Constraints". Give this parameter an initial guess which is equal to the scale factor obtained from the single phase fit. Then constrain the scale factors in the two phases to sum to

100%, using, e.g., @1 and 1.0 − @1 for the scale factors in the `Constraints` tab for the two phases. We want the refinement to start off close to the original single-phase residual minima, so we can give the scale of the original phase a large guess like 0.9 or so. Once this is set up, the remaining parameters can be iteratively refined as before. Be careful not to use too many variables in your fit.

10.6.9.3 *Refinement of conventionally processed films*

You should repeat the refinement steps in Sections 10.6.9.1 and 10.6.9.2 for the PDFs extracted for the conventional TiO₂ films synthesized at 250 °C and 450 °C. Use the same approach to first determine which phases are present in the two samples.

10.6.10   Refining the ITO layer structure

We can additionally try to fit the structure of the nanocrystalline ITO layer, obtained from subtracting the scattering from the glass portion of the substrate, `ito-glass.chi` minus `glass.chi`. As with the TiO₂ PDF data, generate a new fit tree in PDFGUI. Try getting a CIF from PDFITC. For convenience, two possible structures of ITO are also provided with the data: `ITO-bixbyite.cif`, which is of the Mn_2O_3 Bixbyite structure type with cubic space group symmetry $Ia\bar{3}$, and `ITO-Pr7O12.cif`, which is of the Pr_7O_{12} structure type with rhombohedral space group symmetry $R\bar{3}H$.

Load either phase under your fit tree. Start by running the refinement without any parameters to see what it looks like; you can try a fitting range of about 1–15 Å. You can then move forward trying to refine lattice parameters, scale factor, and isotropic thermal parameters.

10.7   Problems

1. How does background subtraction in reciprocal-space compare to subtraction in real-space? Which is preferred?
2. What assumption do we make about the structural relationship between the phases when we subtract the subtrate from the film? How does this affect our interpretation of the resulting signal?
   (a) Identify cases where subtracting a significant background scattering component might not work?
   (b) What are some specific problems you might run into when the background scattering you want to subtract comes from a crystalline component?
3. Describe the residual function resulting from the single-phase fit of anatase to the microwave-assisted TiO₂ film PDF. What clues does this give you about the sample structure?
4. How does the result of the fits to the microwave-grown film compare to those of the conventional films?

5. What did you learn about the structure of the ITO substrate? How does this affect our interpretation of the results of the TiO_2 refinements?

10.8 Solution

10.8.1 How does background subtraction in reciprocal-space compare to subtraction in real-space? Which is preferred?

The difference is that the structure of glass, which accounted for most of the measured signal, was not subtracted out before transforming to the PDF. The resulting PDFs of `tio2-ito-glass.chi` and `ito-glass.chi` then both just *look* like a PDF of the glass substrate. We subtracted the background signal in real-space and hopefully found that the results are the same, once a suitable scale factor could be found (e.g. using the provided optimization code in `background-scale_in_real-space.py`, see Figure 10.4)

The reason for the almost identical PDFs is that the information content is the same in the two cases, just obtained differently. However, finding a suitable background scale can sometimes be more difficult in real-space, depending on the sample. In reciprocal-space, it was in this case easy to visually distinguish the broad amorphous scattering signal of the glass from the small Bragg peaks of TiO_2, which sit on top of it. In fact, we ended up scaling the background signal to line up with the sample signal, except for at the values of Q where the nanocrystalline TiO_2 shows Bragg peaks. Note that if the sample is amorphous or nanostructured and itself only shows a diffuse scattering signal, the background subtraction can be a little more challenging as it is sometimes difficult to distinguish two diffuse signals from each other. In that case, we cannot just line up the background signal with the sample signal as we did here, and it becomes even more important to keep and eye on $F(Q)$ and $G(r)$ during the background scaling.

In real-space, distinct atom-pair correlations exist in both the substrate phases and sample at short distances, and these can sit on top of or near one another, sometimes making the background scaling more difficult when relying only on real-space

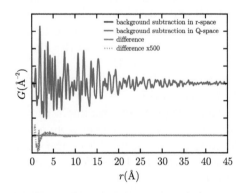

Figure 10.4 Comparison of PDFs resulting from background subtraction in reciprocal-space versus in real-space.

information. There is further room for systematic errors to affect the result when performing the subtraction in real-space because it requires that you process two separate PDFs rather than just one (i.e. you use two separate sets of corrections in reducing the data). In our case, small differences can be seen, for example by zooming very close into the residual of the two: you will see that it is not perfectly zero, especially at very short distances. For these reasons, we generally choose to perform the subtraction with the raw $I(Q)$ data, before reducing and transforming to the PDF, but consider what works best for your system.

10.8.2 What assumption do we make about the structural relationship between the phases when we subtract the substrate from the film? How does this affect our interpretation of the resulting signal?

When performing multiphase analysis from diffraction data, we must often consider what is the interaction between the different components in our sample? Are they randomly and incoherently mixed? Do they coherently interact at their interfaces, such as would be the case for an epitaxial film? Are they intimately mixed such that their individual structures are altered at atomic length scales? In the case of our substrate and film, we are assuming that the phases are two completely separate structures, with negligible interaction at their boundary. Because of this, we can assume that the scattering intensities, or likewise the PDF, for a multiphase sample are a linear combination of the scattering intensities (or PDFs) of the separate phases. We can then subtract away one or more phases, leaving only the phase we wish to analyse. The coefficient in the linear combination is related to the scattering length-weighted intensities of the phases. For a quantitative assessment of the amounts of chemical phases present, make sure to correct the fit coefficients as was discussed in Chapter 5. Care should be taken in making this assessment for other cases.

10.8.3 Identify cases where subtracting a significant background scattering component might not work

As previously mentioned, if the phases are intimately mixed such that their structures are altered at an atomic length scale, then you can no longer subtract either individual phase as a background, simply because that structure no longer exists in your sample. Its scattering intensities (Q-space) or pair-correlations (real-space) are no longer shared with the sample of interest. We also mentioned the case where a coherent interface is formed. In this case, we may still be able to subtract the signal from one of the individual phases, but be on the lookout for signal coming from the structure of the interface, or maybe that is what you are after!

10.8.4 What are some specific problems you might run into when the background scattering you want to subtract comes from a crystalline component?

Liquid and amorphous backgrounds tend to be easier to work with when performing large background subtractions. For crystalline backgrounds, things can become trickier. This can be due to sample processing. Slight inconsistencies between the sample of interest and the background sample can lead to slight errors in the Bragg signal being subtracted. An example of this could be the presence of texturing or coarse particles, which can affect the relative Bragg peak intensities. Even a small change can leave the difference signal unusable. In other cases, very small differences can be mitigated by optimizing the subtraction to minimize any residual features in the PDF left over from un-subtracted signal. In the present case, the ITO subtraction works very well, and we do no see any left-over signal in the PDF. We can think of another general possibility regarding porous framework materials, such as zeolites or metal organic frameworks (MOFs). If we are interested in the structure of some intercalated phase, it is important that the presence of that phase does not cause an expansion of the framework lattice. This will lead to Bragg peaks in your sample of interest shifting from their natural positions in the background measurement!

10.8.5   Describe the residual function resulting from the single-phase fit of anatase to the microwave-assisted TiO₂ film PDF. What clues does this give you about the sample structure?

We can determine the structure of the film by which model gives the best fit or lowest R_w value. Out of the three single-phase refinements performed, you should see that anatase is by far the best. This is particularly the case for the signal at high r, but by inspection of the residual over the whole range, it should be clear that there is an additional signal left over at low r which cannot be accounted for by the peak-sharpening parameters or spdiameter. This discrepancy in the low-r signal intensity can happen for different reasons, most notably for small molecule materials, where the strong bonding within the molecule leads to much sharper peaks than between molecules; however, we typically do not see such a strong effect for oxides. Another case that can occur is that there is some short-range-ordered or amorphous content also present. We can address this case by adding a second phase to the refinement to approximate this disordered component.

If the structural signal left over at low r in the single-phase fits is coming from a second phase, then we should assume that the second phase will have a much lower coherence length or spdiameter, since the fit at high r is already very good. The difference between the structures of anatase, brookite, and rutile at low r is primarily in the orientation and packing of the titanium–oxygen octahedra, which is less different at low r than at high r ,where small differences have a larger distance over which to accumulate and alter the shape of the PDF. Because it is not obvious, we tried all three structures as the second phase along with the best fit structure from the single-phase fit, and we found that that short-range-ordered component actually resembles a more brookite-type packing of octahedra, see Figure 10.5.

10.8.6   How does the result of the fits to the microwave grown film compare to those of the conventional films?

The microwave film was processed at a temperature of 160 °C and a power of 60 W for 1 hour. The resulting structure is nanocrystalline anatase with a

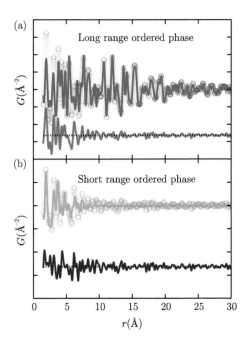

Figure 10.5 Results of structure fitting to the microwave synthesized TiO$_2$ film (blue circles): (a) the PDF from the anatase structure (red) matches best to the long-range-ordered component, leaving a short-range-ordered component (green), which (b) matches best to the PDF from the brookite model (gold). The difference of the second fit is shown (purple).

short-range-ordered component with packing most similar to brookite. Since anatase is the phase we are after, this is great. The conventional film processed at 250 °C ends up quite disordered. In fact, it does not form anatase at all and most closely resembles a brookite-type amorphous packing. The conventional film processed at 450 °C forms a more ordered anatase phase. The results of the conventional film refinements are shown in Figure 10.6. Different processing techniques allowed us to achieve different sample structures and defects or impurities, which may give some capability to optimize between processing cost and product performance.

10.8.7 What did you learn about the structure of the ITO substrate? How does this effect our interpretation of the results of the TiO$_2$ refinements?

Having compared both the Pr$_7$O$_{12}$ and bixbyite structures of ITO to the measurement, you should have found that neither model does a particularly good job of describing the features of the nanocrystalline ITO layer. This should not affect our interpretation of the TiO$_2$ structure, since there is negligible structural coherence between these films allowing us to subtract away the ITO signal. However, the ITO layer is crucial to the formation of the TiO$_2$ layer and for that reason could warrant further investigation.

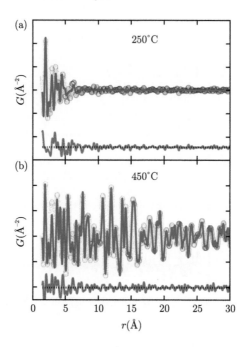

Figure 10.6 Results of structure fitting to the conventionally synthesized TiO_2 films: (a) the film synthesized at 250 °C (blue circles) matches best to the PDF of brookite but only locally ordered, difference (green); (b) the film synthesized at 450 °C (blue circles) matches best to the PDF of anatase (red).

10.9 DIFFPY-CMI solution

This section describes the DIFFPY-CMI solution. It may be skipped on a first reading, but it will be helpful to you when, in the future, you want to learn how to use DIFFPY-CMI. To get started with DIFFPY-CMI, follow the instructions in Section 3.9. The DIFFPY-CMI code for the solution will be in a Python file in the diffpy-cmi folder under solutions for this chapter in your downloaded data. Read through the comments there. They are expanded on below in greater detail where needed.

10.9.1 Running DIFFPY-CMI examples

See the instructions in Chapter 3 for help on running Python files.

10.9.2 fitThinFilm.py

This file presents a DIFFPY-CMI solution to the refinement of several PDF datasets collected from various different thin films, just as you did using PDFGUI above.

To run this code, you should have downloaded the script in a directory called DIFFPY-CMI, with a parent directory called **solutions** and another directory called `data` located at the same directory level as `solutions`. The `data` directory should contain several different subdirectories: `structures`, containing five different `.cif` files, `conventional_films`, and `microwave_film`. The latter two directories should both contain subdirectories named `pdf`, each with a number of PDF files. This example will again take some time to run, as we are running many PDF fits. Have a look at `fitThinFilm.py` and the comments therein. We will explain numbered comments representing significant deviations from the previous DIFFPY-CMI examples.

 # 1–5:   We again begin by importing the necessary packages, giving some details about the measurement, the r-range we'd like to fit over, and initial values for some refined parameters.

 # 6:   In this example we will consider films grown using either microwave or conventional heat-treatment, and we will also attempt to characterize the structure of the ITO layer in one of these films. Here we create a list of strings to identify each of these cases, along with lists of strings for choosing the relevant structure and PDF datasets associated with each of the three cases.

 # 7–45:   Much like the DIFFPY-CMI example in Chapter 8, we will be testing one- and two-phase models against our PDF data. Here we build two functions to make our fit recipe, `make_recipe_one_phase` and `make_recipe_two_phase`, to handle one- and two-phase models, respectively. These functions are defined in much the same way as in Chapter 8, with a few key differences.

We want `make_recipe_one_phase` to be flexible enough to handle both nanocrystalline and bulk materials, so we include an additional boolean argument, `nano`. When `nano` is `True`, we include finite-crystallite size PDF damping; otherwise, we do not. We use an `if...else` statement (points # 13–14) to handle this behaviour. We also choose to not include atomic coordinates in our refinements, and as such we do not loop on the `xyzpars` attribute of the parameter set returned by `constrainAsSpaceGroup` (points # 19+43). Finally, each phase in our two-phase models gets its own damping function, and as such we need to explicitly name the arguments of `sphericalCF` when we register it using `registerFunction` (point # 33). This prevents naming collisions which would occur if we accepted the default argument names.

 # 46–48:   We now begin our `main` function. Like the DIFFPY-CMI example in Chapter 8, we use `glob` to search for PDF data files and structure files in the relevant directories, and we break down the filenames into meaningful strings for use later.

 # 49–51:   We loop first on the different types of films, then on the different PDF data files. We make some decisions about whether the particular film type and datafile are related, and if they aren't we continue to the next iteration of our loop.

 # 52–56: Only the PDF data from films synthesized using microwaves will be treated using two-phase models, so before we enter a loop on all the different structure files, we check if we are working with the microwave film case, and if so, we build a two-phase fit recipe. We build a recipe for each permutation of the anatase phase with one other phase (including itself) by looping on all structure files with TiO_2 in their name.

 # 57–62: We create a unique fit name, including the type of film, the phases we are using, and the name of the PDF datafile. As in all previous examples, we define a list of parameters/tags, free these sequentially and fit, and then write our results to different files.

 # 63: Here we loop over all the structure files we've found, along with their identifying strings which we've parsed out earlier, such that we can consider all relevant phases for our single phase models.

 # 64: If the structure file at the current iteration of our loop is not compatible with the current PDF datafile, we continue to the next iteration of our loop on the structure files.

 # 65–66: We define a number of situations we'd like to consider, represented by permutations of structure files, PDF datasets, and film types, so as to reproduce what we've done using PDFGUI earlier in the chapter. For each of these cases, we define a list of parameters and/or tags we'd like to refine.

 # 67: If we don't match any of these conditions, we continue to the next iteration of our loop.

 # 68–73: If we've made it this far, we have encountered a structure/PDF datafile/film type permutation of interest. We create a unique fit name, including the type of film, the single phase we are using, and the name of the PDF datafile, and we make our fit recipe. We loop over our previously defined list of parameters/tags, free them sequentially and fit, and then write our results to different files.

Once this script has finished, your working directory will be populated with plots of the best fit plots as well as text files describing the fitted parameters and fit statistics for each case considered in your PDFGUI files created in this chapter.

11

Structure of discrete tetrahedral quantum dots: Atomically precise CdSe tetrahedral nanoclusters

Soham Banerjee and Kirsten M. Ø. Jensen

11.1 Introduction and overview

In the previous chapters we have modelled PDFs using models built from crystal structures with periodic long-range order. As we have seen so far, this approach works well not only for bulk, crystalline materials but also for many nanostructured and defective materials, as illustrated in Chapters 8 and 10.

However, in other samples, for example, molecular units, large ionic clusters, or metallic nanoclusters, models based on periodic order are not a good starting point for the modelling. Here we show how PDF modelling can be done without assuming translation symmetry and periodicity. This is done by building up a discrete structural object by specifying atomic coordinates without an assumption of periodicity. The PDF can readily be calculated through such a model using DIFFPY-CMI, for example, using the Debye scattering equation (Debye, 1915) calculator, `DebyePDFGenerator`. There is, at the time of writing, no way to do this with PDFGUI, so this Chapter describes a purely DIFFPY-CMI workflow. Therefore, you need only go through this chapter if you are interested in learning how to use DIFFPY-CMI.

11.2 The question

Semiconductor quantum dots such as CdSe have unique optoelectronic properties that make them useful in a range of applications, such as solar cells and lighting. Their properties are highly dependent on the nanoparticle size, and much effort has therefore been put into mapping the size/morphology/structure/property relations in these systems. We already worked on CdSe in Chapter 8, where we used PDF

Atomic Pair Distribution Function Analysis. Soham Banerjee and Kirsten M. Ø. Jensen, Oxford University Press.
© Simon J. L. Billinge and Kirsten M. Ø. Jensen (2023). DOI: 10.1093/oso/9780198885801.003.0011

(a) $Cd_{35}Se_{20}$ (b) $Cd_{56}Se_{35}$

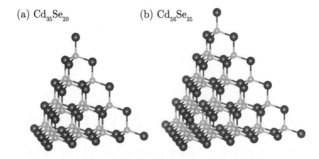

Figure 11.1 Magic-sized tetrahedral CdSe nanoclusters with 35 (a) and 56 (b) Cd atoms.

to characterize stacking faults in nanoparticles. We showed that the stacking fault density is high and is dependent on the particle size. Here, we will work on the structure of even smaller particles. Over the past decade, new synthesis methods have been developed to produce atomically monodisperse, or "magic-sized" CdSe nanoclusters. Such magic-sized clusters can sometimes be crystallized to produce diffraction-quality single crystals, in which case the structure of the material can be solved using single-crystal x-ray diffraction. For example, Beecher *et al* synthesized and crystallized magic-sized CdSe nanoclusters containing just 35 Cd atoms (Beecher *et al.*, 2014). The structure of the clusters was then solved using single crystal diffraction. The clusters have a pyramidal zinc blende structure, as shown in Figure 11.1A. Larger clusters with 56 Cd atoms were also synthesized, but these did not form single crystals. It was hypothesized that their structure was similar to that of the smaller cluster but with extra layers of atoms added to the pyramid surface (Figure 11.1B). Here, the aim is to test if this is the case. We will do that by modelling an experimental PDF with a discrete structure model using DIFFPY-CMI.

11.3 The result

From PDF analysis, it was possible to show that the structure of the $Cd_{56}Se_{35}$ nanocluster can indeed be described as a tetrahedron with the zinc blende structure. The model provided an excellent fit to the PDF. These results support the idea of "quantized" growth of materials on the nanoscale and made it possible to precisely map the relation between size and and structure in CdSe nanoclusters (Beecher *et al.*, 2014).

11.4 The experiment

Experimental conditions are shown in Table 11.1 and the files to download in Table 11.2.

Table 11.1 Experimental conditions.

Facility	NSLS
Beamline	X17A
Detector type	Perkin Elmer amorphous silicon 2D detector
Sample geometry	Powder in 1 mm ID Kapton capillary
Sample environment	100 K, flowing nitrogen crystream
Wavelength	0.31896 Å
Q_{min}, Q_{max}	1.0 Å^{-1}, 20.0 Å^{-1}
Q_{damp}, Q_{broad}	0.06 Å^{-1}, 0.0 Å^{-1}

Table 11.2 Files to download.

Filename	Note
CdSe.gr	Experimental x-ray PDF measured from atomically precise CdSe quantum dots
CdSe.xyz	Atomic coordinates for a 92-atom tetrahedral zinc blende cluster
CdSe-Zincblende.cif	CdSe zinc blende CIF
fitCdSe.py	Python executable to carry out a discrete structure refinement of the tetrahedral
	Zinc blende cluster model and the experimental PDF data

11.5   What next?

1. Download the files and plot the PDF to see what it looks like.
2. Get a CIF file of zinc blende CdSe to use as a starting point for building the structure model.
3. Analyze the measured CdSe PDF in PDFGUI using a crystalline zinc blende structure model and the attenuated crystal approximation.
4. Construct a CdSe structure model by cutting out a tetrahedron from the zinc blende crystal containing 56 Cd atoms.
5. Write a DIFFPY-CMI script to optimize the agreement between the PDF calculated from your discrete zinc blende terahedral cluster (CdSe.xyz) and the measured PDF (CdSe.gr).

11.6   Wait, what? How do I do that?

11.6.1   Download the files

Follow the instructions in Section 3.6.1.

11.6.2   Get a CIF file of zinc blende CdSe to use as a starting point

We provide a CIF with the files for this chapter. You may also see whether uploading the PDF to PDFITC results in STRUCTUREMINING recognizing that the underlying structure of the dots resembles zinc blende and returning the CIF of zinc blende. You can limit the range of r in STRUCTUREMINING, which may affect the result.

11.6.3   Analyse the measured CdSe PDF in PDFGUI using a zinc blende structural model and the attenuated crystal approximation

We start by testing the fit quality we get when we fit the data with a structure model based on a crystalline zinc blende structure, i.e. using a CIF in PDFGUI (or DIFFPY-CMI if you like). We refer to this as an "attenuated crystal refinement" as we simply calculate the PDF from a crystal but attenuate it by refining the sp_diameter parameter. This is the same as was done in the previous chapters on nanoparticle modelling.

   You can follow the steps in Chapter 8 when refining the crystalline zinc blende model to the data. If you have already completed that chapter, you can simply use the same PDFGUI project files and load the new experimental PDF from this chapter (**CdSe.gr**). You can use the same initial parameter values for structural variables that were used in Chapter 8; just remember to update the instrumental parameters (Q_damp and Q_broad) as provided in Table 11.1. You will see that the model actually describes many of the peaks in the PDF quite well, but clear features are seen in the difference curve indicating a non-ideal fit. We want to see if these can be minimized by instead fitting with a discrete model that better reflects the hypothesized structure and shape of the cluster.

11.6.4   Construct a CdSe structure model by cutting out a tetrahedron from the zinc blende crystal containing 56 Cd atoms

We first want to construct a discrete structure model with a tetrahedral shape containing 56 Cd atoms. We thus do not want to describe the structure assuming periodicity but rather create a file with x, y, and z coordinates for each atom in the cluster structure. We usually call this an .xyz file.

   Such a file can be created in several different ways. Using programs such as VESTA (Momma and Izumi, 2011) or CrystalMaker (Palmer, 2015), it is possible to carve out different shapes of clusters from crystal structures by first visualizing more than one unit cell, discarding any symmetry operations, and deleting atoms until you have the shape that you wish to use as a model. The structure that you have carved out can then be exported as an .xyz file. There are also very useful tools that let you create nanoparticles of different structures and shapes in an automated manner.

For example, the Atomic Simulation Environment (ASE) (Larsen *et al.*, 2017) is a very useful set of tools and Python modules that allow setting up, manipulating, visualizing, and analysing atomistic simulations. We have provided a small Python script that uses ASE to build a zinc blende-structured tetrahedron with 56 Cd atoms, and we have used this to create the file `CdSe.xyz` that is provided with the solution files. To run the script, you first need to install ASE. We refer you to the ASE documentation for further instructions and information on the use of the ASE tools.

Using either your own structure model file or the one we have provided, we first want to inspect it visually. If you open the `.xyz` file in a text editor, you will see that it simply consists of a list of atoms with coordinates in Cartesian space. This type of file can be read by most structure viewer programs. Try to open it in, for example, VESTA (Momma and Izumi, 2011) and inspect it – this way you can get a sense of the overall dimensions and geometry of the cluster, make sure the atomic positions are physical (i.e. no overlapping atoms, unexpected dangling bonds, etc.), and also make qualitative comparisons between multiple cluster structures, where applicable.

11.6.5 Write a DIFFPY-CMI script to optimize the agreement between the PDF calculated from your discrete zinc blende terahedral cluster (`CdSe.xyz`) and the measured PDF (`CdSe.gr`)

Refer to earlier sections that describe how to build DIFFPY-CMI scripts, and use example files from the solutions to earlier chapters as a starting point. We provide such a file called CdSe.py with the solutions in case you get stuck.

Run your file to refine parameters to the experimental PDF. You can see the instructions in Chapter 3 for help on running Python files. Here, we just add a few notes on the script provided and the differences between refining a periodic model and a discrete model to a PDF.

 # 1-7: We begin, as we have in previous examples, with the necessary package imports and setting up some global parameters, like the name of our fit, the location of the data and structure files, details about the measurement, and initial values for some fitted parameters. Importantly, some of our parameters in this fit are a bit different, since we are now fitting a discrete model rather than a crystal structure model to the data. For example, we do not refine on any unit cell parameters or fractional atomic coordinates. Instead, we only refine five parameters:

1. The "zoom-scale". This is an expansion or contraction coefficient which, when refined, allows the whole structure to expand or contract isotropically. It serves as a way to fit the interatomic distances without changing the structural model. Its starting value is 1.0 – if it refines to a lower number, the whole structure has contracted, while a larger number means that the structure has expanded in the refinement. It should never be very different from 1.0 as this would mean large changes to the structure model that may not be physical.

2. The scale factor.
3. An isotropic ADP value for Cd.
4. An isotropic ADP value for Se.
5. The δ_2 parameter.

 # 8-9: We create a `make_recipe` function, which can load the structure model and data file.

 # 10-12: Adopting the same approach as in previous examples, we create a profile generator and a fit contribution, and we give the latter our experimental profile and our PDF profile generator. Note that here we use the `DebyePDFGenerator` to calculate our PDF rather than `PDFGenerator` which we have used in previous chapters. While the `PDFGenerator` can calculate PDFs from crystal structures, the `DebyePDFGenerator` is used for calculations of PDFs from discrete structures. As the name implies, the `DebyePDFGenerator` uses the Debye equation (Debye, 1915) for calculating a scattering pattern and subsequently obtains a PDF by Fourier transforming the resulting $F(Q)$.

 # 13: We set an equation within the `Fit Contribution`. Here we simply have one Generator, "G1", and a scale variable "s1".

 # 14: Finally we create a fit recipe object which holds all the details of the fit. We assign our fit contribution to the new fit recipe through its `addContribution` method.

 # 15: We initialize the instrument parameters (Q_{damp} and Q_{broad}) and assign Q_{min} and Q_{max}. These are kept fixed in the refinement.

 # 16-19: We add the five variables and initialize them.

The rest of the steps in the script should be familiar by now; if not, refer to the previous chapters, and especially Chapter 3.

You can now compare the quality between the fits made with the crystal structure model and the discrete model.

11.7 Problems

1. What are the differences between calculating PDFs within the attenuated crystal (AC) approximation, as discussed in Chapter 8, and the discrete models used here?
2. Were the AC fits good enough? What are some motivating factors for testing the discrete cluster models?
3. What are some challenges when using discrete structure models?

11.8 Solution

Table 11.3 and Table 11.4 summarize what we got from modelling with the attenuated zinc blende crystal model and the discrete CdSe structure.

The discrete zinc blende model clearly gives a much better fit to the experimental PDF than the attenuated crystal structure. This is seen in Figure 11.2, where the two fits are compared. For example, it describes the broad PDF peak at ca. 15 Å, which the crystal structure model did not catch. This happens because the crystal structure model assumes a spherical particle and refines its diameter to 17 Å. The features at the highest r-values in the experimental PDF relate to the distances between atoms in the corners of the tetrahedron and are thus not taken into account when using a spherical particle model. The tetrahedral geometry of the cluster also means that the peaks between 4 and 10 Å are much better described than with the attenuated crystal model.

The zoom-scale parameter refines to 0.99. Thus, we see a small contraction of the interatomic distances compared to the starting model. When constructing the `.xyz` file in ASE, we used a lattice constant for the zinc blende structure of 6.1 Å. Therefore, a small contraction agrees well with the results from the attenuated

Table 11.3 Fit results using the attenuated crystal structure model.

R_w	0.21
Scale factor	0.312
a (Å)	6.050
$U_{iso,Cd}$ (Å^2)	0.028
$U_{iso,Se}$ (Å^2)	0.055
δ_2 (Å^2)	5.74
sp_{diameter} (Å)	17

Table 11.4 Fit results using the discrete cluster model.

R_w	0.103
Scale factor	0.352
$U_{iso,Cd}$ (Å^2)	0.0437
$U_{iso,Se}$ (Å^2)	0.0178
δ_2 (Å^2)	5.33
zoom-scale	0.990

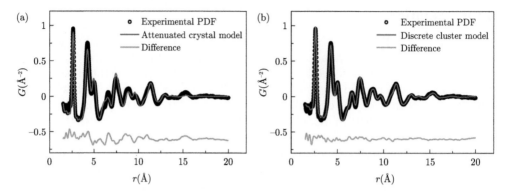

Figure 11.2 a: Fit to the experimental PDF using the attenuated crystal model. b: Fit to the experimental PDF using a discrete cluster structure model.

crystal model, where the lattice constant a refines to 6.05 Å. The refined ADPs show somewhat different values – for the first refinement (attenuated crystal), Cd has a higher ADP value than Se, while it is the opposite for the discrete structure modelling. This mostly indicates that there is some correlation between the two parameters and that we should be careful when interpreting the ADP values. For example, we could try to make a new refinement where the two ADP variables are constrained to take the same value and see how that affects the resulting fit.

Apart from small nanoclusters as described here, the use of discrete cluster modelling is also very useful for structural analysis of molecular units in organic materials, large ionic clusters, or even the local structure of amorphous materials. The challenge is often coming up with a good starting model for refinements. Compared to crystalline or nanocrystalline materials, where a starting model can sometimes be found by considering the presence of specific Bragg peaks, it can be more difficult to find out where to start when modelling the structure of non-periodic objects. Often, it is useful to still consider the structure of related crystalline materials and, as done here, carve out different structures and shapes which can be tested against the data. In such cases, good knowledge of the structural chemistry of the system in question is important. For example, in the example in this chapter, the model was developed based on knowledge of smaller, but closely related clusters. In other cases, information from DFT simulations may be useful. Furthermore, some automated tools have been developed for identifying starting structure models. For metallic clusters, "cluster mining" can be used to quickly test a range of different cluster structures from a cluster library to automatically find the model that provides the best agreement with the data (Banerjee *et al.*, 2020). Methods for automatically testing different fragments of various structures have also been developed, which can help in identifying which structural motifs may dominate in a sample (Lindahl Christiansen *et al.*, 2020b; Anker *et al.*, 2022).

In the refinements carried out above, we refined very few variables: a scale factor, the expansion/contraction factor, and the ADPs of Cd and Se. This strategy is often used in discrete structure modelling. We could in principle also refine the atomic positions. However, compared to refinements using crystal structure models where space group symmetry can be applied, the atoms in the discrete model are all considered independent, and the number of variables can increase quickly unless you introduce chemically reasonable constraints, for example by introducing sensible symmetry constraints based on the nature of the structure model. Such constraints can be implemented in DIFFPY-CMI by carefully considering how different atoms should move in relation to each other.

With this chapter, we hope to have provided you with the tools needed to start analysing PDFs using discrete modelling. As already mentioned, this approach can be useful for a range of different projects and samples. There are many examples of this in the literature, and we refer the reader to reviews on nanoparticle analysis for further inspiration for modelling (Lindahl Christiansen *et al.*, 2020a).

12

Structure and intercalation environment of disordered layered materials: Zirconium phosphonate-phosphate unconventional MOFs

Maxwell W. Terban

12.1 Introduction

Layered materials are relevant to technologies within a wide variety of applications, such as energy and environmental remediation. Such materials are built up of distinct, layered motifs, and variations in the stacking relationship between the layers can lead to different types of disorder in the material. It is important to be able to characterize the disorder as it will have large influence on the material properties.

Many types of layered materials exist. In some cases, the bonding strength is similar both within and between the layers, and variations in stacking are allowed by the existence of equivalently favourable positions in which a layer can sit with respect to its neighbouring layers. This occurs, for instance, in close-packed, monatomic compounds, which can exist as either fcc (ABCABC stacking) or hcp (ABABAB stacking). Defects such as twin boundaries can lead to intermediate stacking states such as ABCBA, which lead to differences in the distribution of interlayer atom-pair distances (Banerjee *et al.*, 2018). Similar scenarios can occur in binary compounds. For example, stacking faults between wurtzite- and zinc blende-type structures were discussed in Chapter 8 for CdSe nanoparticles.

In other cases, the interactions between layers can be much less well-defined, and further types of interlayer disorder can develop, such as turbostratic disorder, where the orientational ordering between layers can be significantly reduced or lost completely. This can occur, for example, in layered carbons (Petkov *et al.*, 1999), layered double hydroxides (Funnell *et al.*, 2014), so-called MXenes (Anasori *et al.*, 2016),

Atomic Pair Distribution Function Analysis. Maxwell W. Terban, Oxford University Press.
© Simon J. L. Billinge and Kirsten M. Ø. Jensen (2023). DOI: 10.1093/oso/9780198885801.003.0012

and polymerized carbon nitrides (Schlomberg *et al.*, 2019). Various modelling schemes have been developed for different cases. For example, it is possible to generate complex many-layer models from which scattering patterns can be simulated to investigate the full effect of stacking disorder in reciprocal-space (Treacy *et al.*, 1991; Bette *et al.*, 2017; Metz *et al.*, 2018). However, in some cases, assumptions can be made that allow much simpler single-layer models rather than the full many-layer statistical models to yield scientific insights (Farrow *et al.*, 2013).

In this chapter, we investigate a zirconium phosphonate-phosphate unconventional metal organic framework (UMOF) consisting of zirconium phosphate-type layers that are interlinked by phenyl biphosphonate groups. The structure of the UMOF material can be related to that of crystalline zirconium phosphate, here denoted α-ZrP. This structure is illustrated in Figure 12.1(a) and consists of layers of corner-shared ZrO_6 octahedra and PO_4 tetrahedra. Water molecules then intercalate between the layers. On the other hand, the layers in the UMOF materials are separated by organic linkers as illustrated in Figure 12.1(b). The weak interactions between the inorganic layers and partial occupancy of the interlinking phenyl groups allow for significant orientational disorder from one layer to the next. This disorder is so significant that there is little signal from interlayer atom-pair correlations in the scattering data from such materials, which allows us to use one single layer motif as an approximate model for our analysis. In doing so, we can investigate the structure of the inorganic layer and study how different ions or molecules can be

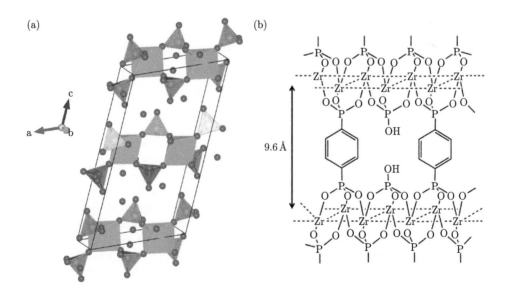

Figure 12.1 a) Structure of crystalline α-ZrP. Zr is shown in green, phosphorous in purple, and oxygen in red. Hydrogens have been omitted. The structure is visualized using VESTA (Silbernagel *et al.*, 2016a). b) Suggested portion of the zirconium phosphonate-phosphate hybrid structure. (b) Adapted with permission from Silbernagel *et al.*(2016a). Copyright 2016 American Chemical Society.

incorporated between, and interact with, the layers. In this chapter we need both PDFGUI and DIFFPY-CMI for modelling.

12.2   The question

In this study, we seek to characterize the structure of the zirconium phenyl-phosphonate-phosphate UMOF. Due to the significant disorder in the material, answering this question is not feasible from x-ray powder diffraction patterns alone, and we therefore use both qualitative assessment and quantitative PDF modelling to determine the structure. The first goal is to understand the structure of the layers of the two UMOFs denoted as "H-Zr" due to the protonated state of the open phosphate groups. The two UMOF samples have different stoichiometries, with phenyl-biphosphonate linker to phosphate (PP) ratios of 2:1 (H-Zr (2:1)) and 1:0 (H-Zr (1:0)), both prepared using phosphoric acid as the phosphate source.

Having a sound picture of the UMOF structure, the goal is then to characterize the UMOF (H-Zr (2:1)) loaded with Tb^{3+}. UMOF materials like this are useful ion exchangers, which are promising for separating lanthanide and actinide ions in order to recycle nuclear fuel waste (Silbernagel *et al.*, 2016a; Silbernagel *et al.*, 2016b). It is important to determine the local and long-range structures in order to better understand the mechanism underlying their preference for higher-valence intercalated ions. To this end, we again revisit the difference PDF (dPDF) method introduced in Chapter 10 to investigate the local environment of Tb^{3+} ions intercalated between the layers by comparing against the local environment of various terbium phosphate materials.

12.3   The result

We found that the structure of the inorganic layers in the zirconium phenyl-phosphonate-phosphate UMOF is analogous to the layers in α-ZrP. The large interlayer spacing between the zirconium phosphonate layers is dictated by the length of the phenyl linkers, and there is a significant loss of discernible ordering between neighbouring zirconium phosphate layers. The absence of linkers at 50% of available sites contributes to this lack of order. However, local density variation between layers is observable and becomes increasingly ordered when the ratio of phosphonate to phosphate sites is increased. The dPDF analysis showed that Tb^{3+} ions are taken up by the H-Zr ion exchange material and prefer to maximize coordination to negatively charged O atoms dangling from the phosphate groups of the inorganic layers. By comparison to experimental and theoretical Tb phosphate materials, the local environment of the Tb^{3+} ions between the layers is best represented by that in a distorted version of a hypothetical Tb phosphate Scheelite-type structure. All of this is discussed in detail in Terban *et al.* (2017).

12.4 The experiment

Please see Table 12.1 for the experimental information and Table 12.2 for the files to download.

Table 12.1 Experimental conditions.

Facility	NSLS-II
Beamline	XPD
Detector type	Perkin Elmer amorphous silicon 2D detector
Sample geometry	Powder in 1 mm ID Kapton capillary
Sample environment	100 K
Q_{damp}	0.0382 Å^{-1}
Q_{broad}	0.0192 Å^{-1}

12.5 What next?

1. Download the files.
2. Plot and compare the total scattering data for samples α-ZrP and H-Zr (2:1). Note their similarities and differences. Do the same for samples H-Zr (2:1) and H-Zr (1:0) and for samples H-Zr (2:1) and H-Zr (2:1) with Tb.
3. Use PDFGETX3 or xPDFSUITE to generate PDFs from the datasets.
   (a) Load the data `aZrP.chi` and subtract `kapton.chi` as a background to extract the $I(Q)$, $S(Q)$, $F(Q)$, and $G(r)$ for α-ZrP (the background scaling should be 1.0 and the composition is "Zr H2 P2 O8 H2 O").
   (b) Do the same as above for the `H-Zr_1-0_PP.chi` and `H-Zr_2-1_PP.chi` data. Use the composition: "Zr0.34 O1.02 P0.34 C2.04 H1.36 P0.34 O1.02 O3 P O H O0.64 H0.64 H2.9 O1.45".
   (c) To best assess the structure of the inorganic layer component, we can remove obvious interlayer correlations (which will show up as long-wavelength oscillations in the PDF due to low-Q Bragg reflections) from the UMOF data before the Fourier transformation. Reprocess the previous datasets with the 002 Bragg peak (the first peak in the diffraction pattern) removed from the Q-range used in the Fourier transformation. Save the resulting files using a different output name.
   (d) Replot and compare the H-Zr PDFs with and without the first Bragg peak included.
   (e) Exit PDFGETX3/xPDFSUITE.
4. Get a CIF file from PDFITC to use as a starting point.
5. Use PDFGUI to refine and confirm the crystal structure `alpha_ZrP.cif` for the α-ZrP sample.

Table 12.2 Files to download.

Filename	Note
Data	
`aZrP.chi`	`.chi` file of α-ZrP
`H-Zr_2-1_PP.chi`	`.chi` of H-Zr with 2:1 PP ratio
`H-Zr_2-1_PP_withTb.chi`	`.chi` file of H-Zr (2:1) loaded with Tb
`H-Zr_1-0_PP.chi`	`.chi` file of H-Zr with 1:0 PP ratio
`kapton.chi`	`.chi` file from background measurement
`aZrP.gr`	Target PDF for α-ZrP
`H-Zr_2-1_PP_cutFirstPeak.gr`	Target PDF for H-Zr 2:1 PP w/o first Bragg peak
`H-Zr_2-1_PP_withFirstPeak.gr`	Target PDF for H-Zr 2:1 PP w/ first Bragg peak
`H-Zr_1-0_PP_cutFirstPeak.gr`	Target PDF for H-Zr 1:0 PP w/o first Bragg peak
`H-Zr_1-0_PP_withFirstPeak.gr`	Target PDF for H-Zr 1:0 PP w/ first Bragg peak
`H-Zr_2-1_PP_TbDifferencePDF.gr`	Target dPDF for H-Zr (2:1) loaded with Tb
Structures	
`alpha_ZrP.cif`	Structure of α-ZrP (Clearfield and Smith, 1969)
`discrete_alpha_layer.xyz`	`.xyz` format structure of single layer of α-ZrP
`unit_cell_layer.xyz`	`.xyz` format structure of single unit cell layer
`alpha_ZrP_enlargedZ.cif`	α-ZrP structure with enlarged c dimension
`tbpo4_zircon-type.cif`	Structure of zircon-type terbium phosphate
`tbpo4_scheelite-type.cif`	Structure of scheelite-type terbium phosphate
Code	
`discrete_debye_calculator.py`	Simulate PDF from a discrete layer
`cart_to_frac_coordinates.py`	Place Cartesian atomic coordinates in a new unit cell

6. Try fitting the same structure to the H-Zr (2:1) sample.
7. Use DIFFPY-CMI to simulate a PDF for a single layer of α-ZrP using the Debye PDF calculator, and compare to the H-Zr (2:1) PDF (with the first Bragg reflection removed).
8. Use DIFFPY-CMI to further refine the single-layer model. Note, PDFGUI does not support fitting with discrete structures. Try the same for H-Zr (1:0).
9. Use PDFGETX3 or xPDFSUITE to obtain a dPDF for the terbium correlations by subtracting `H-Zr_2-1_PP.chi` from `H-Zr_2-1_PP_withTb.chi`.
10. Identify pairs giving rise to the peaks in the dPDF using the supplied terbium phosphate structures.
11. Use PDFGUI to fit dPDFs from the terbium phosphate structure, and determine the most similar local environment of the Tb^{3+} ions.

12.6 Wait, what? How do I do that?

12.6.1 Download the files

Follow the instructions in Section 3.6.1.

12.6.2 Get a CIF file from PDFITC to use as a starting point

Using STRUCTUREMINING in PDFITC is a great way to get CIF files to make a start on your refinement campaign. Please see Section 6.6.2 for more details. For convenience, we also provide the relevant CIFs with the download files.

12.6.3 Plot the total scattering data from the four samples

Plotting and comparing your total scattering data is an important sanity check before you start generating PDFs. Plot the data from α-ZrP and from one of the UMOFs, H-Zr with 2:1 phosphonate:phosphate ratio files using your preferred plotting program. See Section 4.4 for help with `.chi`, `.iq`, etc., file formats. See Section 4.6.2 for help with doing this using Diffpy tools. Note that the scattering intensity is given as a function of Q in nm^{-1}. Notice that while the crystalline α-ZrP sample shows distinct Bragg peaks, the pattern from the H-Zr sample has broader features. In fact, broad, triangular-shaped peak profiles are often an indicator of turbostratic disorder (Warren 1941). When comparing the two UMOF samples, H-Zr 2:1 and H-Zr 1:0, we see similar features in the two scattering patterns, but the relative intensities of some of the peaks differ between the samples, primarily the first Bragg peak in the pattern, which can be related to the interlayer ordering. When comparing the H-Zr 2:1 sample with and without Tb^{3+}, we again see similar features, but the Q-dependence of the scattering intensity has changed. We will use PDF to understand these changes.

12.6.4 Process data with PDFGETX3/XPDFSUITE to obtain PDFs

We start by obtaining PDFs from the three samples without Tb^{3+}: α-ZrP, H-Zr 2:1, and H-Zr 1:0. Use either PDFGETX3 or XPDFSUITE to obtain PDFs from the data. Refer back to Chapter 4 if you need a reminder on how to use the programs.

Data from all samples were collected in Kapton tubes. Use the file `kapton.chi` for background subtraction. For all three samples, set the background scale to 1.0. The composition for α-ZrP is "Zr H2 P2 O8 H2 O", and for the hybrid UMOF sample, you can set the composition to "Zr0.34 O1.02 P0.34 C2.04 H1.36 P0.34 O1.02 O3 P O H O0.64 H0.64 H2.9 O1.45". Note that this is not exactly the composition of the H-Zr 1:0 sample, but it is sufficient for the sake of comparison as the effects due to the phosphonate ratio in the composition are negligible. The data quality is good over the whole range measured. However, since we are most interested in confirming the long-range structure of the hybrid UMOF, it is useful to choose a lower Q_{max} such that the high-frequency noise ripples are minimized around the highest-r observable peaks. In the study published, we used a Q_{max} of 18.0 Å^{-1}. It is advised to choose an r_{max} value which is higher than the distance where the signal is completely lost due to damping from the finite experimental resolution, e.g. 100 Å. It is quick to compute the measured PDF over a longer distance now and avoid the possibility of having to come back and redo the calculation later because you want to see the behaviour at high r.

You can compare your PDFs to those that we provided, and you may use either for the subsequent modelling part. The exact values used in the analyses can be found in the header information of the `.gr` files.

12.6.5 Reprocess data from UMOF samples to isolate the intralayer correlations

Later in this chapter, you will investigate the UMOF structure using a 2D, single-layer model. Using this assumption, you can minimize the effects of the interlayer correlations in the resulting PDFs. While there is no way to completely separate out all possible scattering contributions from between layers, the large interlayer density modulations can at least be partially removed by cutting the first strong Bragg reflection 002 at $Q = 0.66$ Å (see Figure 12.2). The 002 peak arises from layer periodicity, and its effect can be observed in the PDF from the H-Zr (1:0) as a long-wavelength sine wave. If you look at this oscillation at high-r distances and simultaneously increase the Q_{min}, you will see that the oscillation disappears and becomes completely flat just after the Q_{min} rises high enough to completely cut off the 002 reflection from the Q-range used in the Fourier transformation. Remember that this does not remove all possible interlayer contributions in the PDF. However, in compounds like these, we do not expect significant interlayer contributions with long-range order, and therefore the resulting PDF will be dominated at high distances by the structure of the layer alone. For these data, we used a Q_{min} value of

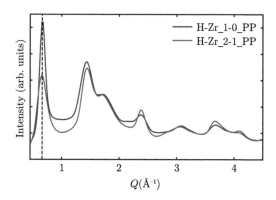

Figure 12.2 Comparison of the scattering intensities for the H-Zr_1-0_PP and H-Zr_2-1_PP samples. The dashed line at 0.66 Å indicates the position of the interlayer 002 peak.

0.78 Å^{-1}. Save the PDFs from the two UMOF samples with the new $Q_{\min}$ parameter under a different name.

12.6.6   Compare the PDFs from the three samples

At this point, it is a good idea to plot your PDFs and consider similarities and differences. For example, if you compare PDFs from the three samples and consider the low-r region, you will see that the peaks from the UMOF PDFs almost completely overlap with the peaks from the crystalline α-ZrP. This shows us that the local structure of the UMOF layers is closely related to that of α-ZrP despite the lack of long-range order. We will use this information when we start modelling.

12.6.7   Start a PDFGUI project and start a new fit

Refer to Chapters 3 and 6 if you need help getting started in PDFGUI.

12.6.8   Refining the α-ZrP structure

12.6.8.1 α-ZrP

We will start by modelling the PDF from the crystalline α-ZrP sample. Insert the dataset into your fit tree, using the .gr file resulting from aZrP.chi. Obtain a CIF file using PDFITC, or you may use the alpha_ZrP.cif file we provided, and upload it as a phase in the fit. Under the data set Configure tab, input the given experimental resolution parameters, Q_{damp} and Q_{broad}, and set your refinement range to something like 1.0–35.0 Å. Before refining any structure parameters, you can first calculate the PDF of the phase for initial comparison without any constraints set by clicking on the gear in the top left of the PDFGUI window. Inspect the result.

You can then proceed by incorporating simple structure parameters such as the lattice parameters (taking into account the monoclinic symmetry), i.e. a, b, c, β, and a scale factor. Run the refinement again, and copy the refined values to the initial column. Proceed further by adding isotropic atomic displacement parameters (ADPs), the sp_{diameter} variable, and a δ_1 or δ_2 parameter to account for correlated motion. You could go further and try refining atomic sites by symmetry; however, we advise not to do this for more complicated structures with low symmetry and many atoms as atoms may refine to non-physical positions without constraints. You can, as always, see how we have set up our refinements in the Solutions folder distributed with the data.

12.6.8.2 H-Zr (2:1)

The same procedure can be performed for the `.gr` file resulting from `H-Zr_2-1_PP.chi` with the 002 peak included. The resulting goodness-of-fit should be quite poor. In this case, you can try refining the Zr and P atom positions taking symmetry constraints into account. However, this does not improve the fit, and this observation helps to further verify that the differences in the structures are not simply due to a slight distortion or rearrangement of the atoms within otherwise-ordered layers.

12.6.8.3 Simulating the PDF for a discrete single layer

Since we cannot model the PDFs from the UMOF samples with the known crystal structure for α-ZrP, we need to try something different. We know from above that the α-ZrP and UMOF structures are closely related but that there is probably interlayer disorder in the UMOF samples. We will therefore try to simulate a PDF from *one* zirconium phosphate layer and see how that compares to the PDFs we obtained from the UMOFs. PDFGUI does not support PDF analysis using discrete structure models, though there are workarounds involving the use (or abuse, shall we say) of PDFGUI that we will discuss later in this chapter. Here we continue using DIFFPY-CMI.

First, you need to obtain a discrete structure model of a single layer from the α-ZrP structure in `alpha_ZrP.cif`. This is done by cutting out a discrete layer from the α-ZrP structure, which can be accomplished using most structure visualization packages, such as CrystalMaker, Mercury, or VESTA. Here, we have provided the model in the file `discrete_alpha_layer.xyz`, which you can open in any of the visualization packages mentioned. We recommend that you open this and explore the structure so that you know what you are modelling. The PDF for a discrete structure model can be easily calculated using the calculators in DIFFPY-CMI. Here we use the Debye PDF calculator in order to properly account for and remove the small-angle scattering signal. You can inspect and run the code provided in `discrete_debye_calculator.py` to simulate the PDF for the discrete structure provided. In the script, you need to tune the simulation parameters for the structure, such as element specific ADPs or δ_2, and for the experiment, such as Q_{max}, Q_{min}, Q_{broad}, and Q_{damp}. Now consider the PDF that results. How does this compare to the experimental PDF obtained from the UMOF structure? Note that it is extremely important to give a non-zero $Q_{\min}$ value, typically around 0.5–1.5, as the scattering effects at low-Q simulated from the model in this case are not representative of the real measured data.

12.6.9 Refining the single layer structure

Having established that a single-layer structure may be a good starting point for the UMOF PDF modelling, we now proceed to consider how we can refine the structural parameters to get a better description of the experimental PDFs.

There are two ways in which you can consider fitting a layer model to the measured data. As we saw above, discrete structures can be cut out from the bulk crystal structure and saved as `.xyz` files. Such discrete models of atomic structure can be very useful in certain cases, such as discrete nanoclusters as we saw in Chapter 11. However, as the number of atoms increases, running the refinements becomes quite slow. Also, a large number of different structures, or slabs, must be prepared, and separate refinements must be run to test different ranges of structural coherence.

A simpler way to perform the refinement of a 2D layer is to isolate the single layer as it is already defined within the crystalline unit cell and keep translational symmetry. In order to remove all interlayer correlations, other layers within the unit cell must be removed. However, if you kept the original unit cell size, you would still see interlayer peaks in the PDF arising from pairs of atoms between the primary layer and the same layer in neighbouring unit cells due to translational symmetry along c. Thus, the c unit cell dimension (i.e. the unit cell vector along the stacking direction) must be expanded so that that the distance between neighbouring layers is larger than the largest distance over which the structure is refined to the PDF. The idea behind this trick is that, although interlayer correlations (interatomic distances) will be present in the periodic PDFGUI model, if we separate the layers by a distance larger than the r-range that we are fitting over, these interlayer peaks will not appear in the calculated model, which will therefore consist only of intralayer correlations. Unfortunately, this will result in the wrong atomic density in the model, and the resulting baseline will thus be incorrect. This necessitates the use of DIFFPY-CMI as discussed further below. We briefly describe the steps needed using Python code.

It is more complicated than simply expanding the c-axis dimension because in the PDFGUI model, the atomic positions are defined in fractional coordinates $[xyz]$, and the atoms in the target layer would then move away from each other during the c expansion. We therefore need to find new fractional coordinates in the expanded cell that preserve the bond-lengths within the layer. To do this we first extract atoms of a single layer in the unit cell and convert them from fractional to Cartesian coordinates (here denoted by a $'$ symbol) $[x'y'z']$. The first step is therefore to use any structure visualization software to save a one-unit-cell section of a layer as a `.xyz` file, which can be saved directly as Cartesian coordinates $[x'y'z']$. You can use the provided `.cif` to make this file and check it against the `unit_cell_layer.xyz` file that we made to make sure you have the correct segment. When you have this layer, the atoms can be placed into a new unit cell with the expanded c using the following matrix transformation, which produces the new fractional $[xyz]$ coordinates from the Cartesian coordinates $[x'y'z']$:

$$
\begin{bmatrix} x \\ y \\ z \end{bmatrix} = \begin{bmatrix} \frac{1}{a} & -\frac{\cos(\gamma)}{a\sin(\gamma)} & bc\frac{\cos(\alpha)\cos(\gamma)-\cos(\beta)}{V\sin(\gamma)} \\ 0 & \frac{1}{b\sin(\gamma)} & ac\frac{\cos(\beta)\cos(\gamma)-\cos(\alpha)}{V\sin(\gamma)} \\ 0 & 0 & ab\frac{\sin(\gamma)}{V} \end{bmatrix} \begin{bmatrix} x' \\ y' \\ z' \end{bmatrix}
\tag{12.1}
$$

where V is the volume of the new cell, which can be computed by,

$$
V = abc\sqrt{1 - \cos^2(\alpha) - \cos^2(\beta) - \cos^2(\gamma) + 2\cos(\alpha)\cos(\beta)\cos(\gamma)}.
\tag{12.2}
$$

This is a general relationship for any unit cell, but you can quickly see how this expression simplifies as you increase the symmetry of the cell (for example, what happens to the matrix elements when some unit cell angles are 90°?).

A script to perform this transformation can be easily written in Python; we have provided this. You can inspect and run `cart_to_frac_coordinates.py` to see how this is performed. The output will give a list consisting of the atom identifier, the atom type, and its new fractional coordinates x, y, z.

The new coordinates can then be easily placed back into a `.cif` file with the updated unit cell values, using the original `α-Zr.cif` as a template. We have provided our new file, `alpha_ZrP_enlargedZ.cif`, with the data.

Once this structure transformation is done, the new *single-layer* structure can be refined using DIFFPY-CMI on the `H-Zr_2-1_PP.gr` and `H-Zr_1-0_PP.gr` datasets obtained with the first Bragg peak cut off. Note that, because of the fictitious interlayer spacing in our new model, the model does not have the correct atomic density. Since the PDF baseline depends on this density, a PDF calculated with a real-space calculator (as PDFGUI uses) will underestimate the downward deviation of the baseline. There are tricks that can be played to fix this, for example, placing a fake atom in the structure with a large mass but low scattering factor and large ADP, but we won't do that here. Rather, we will use the Debye PDF calculator in DIFFPY-CMI. To get a good fit, we must ensure that the $Q_{\min}$ value is high enough to cut off any spurious diffraction effects from the out-of-plane expanded layer periodicity that is present in the model but not in the sample. A script to do this is provided in `refine_H-Zr_2-1.py`. Try running the refinement against the PDFs obtained with and without the first Bragg peak included in the Q-range. Note the differences in the fit quality and features in the residual. It it important to note that this simplified model is simply one way to approximate our real material to get a better understanding of the structuring, and if we want to further predict properties from the structure, we cannot neglect correlation between neighbour layers or the interactions of the phenyl groups, for instance. It is also worth noting that despite the fact that we extracted a layer from a crystal structure with $P21/c$ symmetry, the atomic site positions in the real H-Zr structure do not necessarily obey the same site symmetry restrictions, though this is also a suitable approximation to avoid overparameterizing our model. You can see what effects the sample composition has on the goodness-of-fit and consider what this might mean in terms of the strength of interlayer interactions and impact on the structure.

12.6.10 Determining the dPDF for Tb with PDFGETX3/XPDFSUITE

In the previous sections, we have developed a model to describe the essential structure components of the zirconium phosphate layers in these UMOFs. Such a model allows us to better understand the structure/property relations in the materials, as we will discuss more in the Solutions section below. For now, the next step is to treat the data from the Tb^{3+}-loaded UMOF to find out how and where the terbium ions sit in the structure.

To do so, we will generate a dPDF, where we isolate PDF peaks arising from the interaction between Tb^{3+} and the zirconium phosphate layers by subtracting the signal of the unloaded UMOF from that of the loaded UMOF. A dPDF can be obtained by subtraction in reciprocal- or in real-space, as discussed in Chapter 10. Here, we will do the subtraction in reciprocal-space, and you should obtain the difference scattering signal by subtracting the scattering of the `H-Zr_2-1_PP.chi` sample from `H-Zr_2-1_PP_withTb.chi`. Since the overall structure of the framework does not change significantly from the Tb^{3+} loading, this should give a reasonable approximation of the coherent scattering signal from interferences between Tb atoms in the loaded framework with the other P, O, C, and Tb atoms in the structure, from which we can obtain the dPDF.

You can do this subtraction directly in PDFGETX3 or XPDFSUITE by using the data from the unloaded UMOF as the background file. As discussed in Section 10.6.3, when the background and sample signals are comparable, as here, the background subtraction requires care. The background scale is adjusted to optimize some particular signal in the dPDF. Here we monitor and minimize peak intensities which come from specific non-Tb intralayer distances, for example P–O at 1.54 Å, Zr–O at 2.06 Å, or Zr–Zr at 5.33 Å. The best dPDF can be obtained by minimizing the intensity of these peaks. Using this method, we found that a scale factor of 0.81 gives an optimal dPDF. You will observe a shift of the peak at 2.06 Å to 2.37 Å, which is also a good indicator that you are now observing expanded Tb–O distances and not just residual un-subtracted Zr–O signal.

12.6.11 Indexing distances in the dPDF

Having obtained a good dPDF, we can start analysing it to get information on the structure of the Tb^{3+}-loaded UMOF. A useful way to analyse the difference signal is simply to index the PDF peaks, i.e. to try to identify which relevant atomic pairs could give rise to a peak at the specific r-value. For nearest-neighbour distances, candidates can be found by referring to tables of ionic radii, for example, the Shannon tables (Shannon 1976). These show that Tb^{3+} to O^{2-} distances would be 2.273 Å for VI-coordinated Tb and 2.39 for VIII-coordinated Tb using the value for II-coordinated O. In Figure 12.3 we indeed see a strong peak at approximately 2.37 Å.

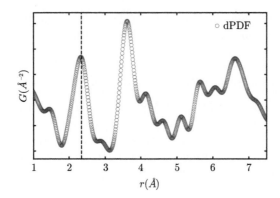

Figure 12.3 Plot of the difference extracted between the H-Zr_2-1_PP samples with and without Tb loading. The dashed line at 2.37 Å indicates the position of the probable Tb–O pair distance.

The next step is to do a comparison to PDFs calculated from known structures. Structure files can be mined from existing databases, for example by uploading the dPDF to STRUCTUREMINING at PDFITC and looking for compounds that contain Tb, P, O, and anything else, or doing it manually by exploring databases such as ICSD, WebCSD, COD, or The Materials Project. In the current case, since the work was done before PDFITC existed, all structures containing any Tb phosphate motifs were extracted from the ICSD and The Materials Project, and the interatomic distances in these structures were extracted to compare with the experimental PDF in order to "index" the peaks.

Information about element-specific atomic distances in a structure model can be easily extracted for this purpose using tools from both PDFGUI or XPDFSUITE. In PDFGUI, first select a structure (i.e. a `.cif` loaded under the fit tree). Then, at the top of the screen, select **Phases**, and then **Calculate bond lengths**. A new window will open, in which you can select the atomic pairs for which you want to find the interatomic distance. Here, we choose **Elements: Tb** and **Elements: All**, and a range, for example 1–10 Å. The resulting bond lengths will appear in the PDFfit2 Output dialogue in your main PDFGUI window. In XPDFSUITE, the distances can be extracted by selecting the ruler button from the upper tab. Then choose a structure file to import, and give input on the relevant atomic pairs and r-range.

We have provided two `.cif` files for related materials: a zircon-type terbium phosphate structure `terbium_phosphate_zircon-type.cif` and a hypothetical Scheelite-type terbium phosphate structure `terbium_phosphate_scheelite-type.cif`. Try the different methods described above, and identify the atom-pair contributions to the peaks in the dPDF.

12.6.12 Fitting the dPDF

From the structure fingerprinting procedure described above, we should be able to identify the features in the dPDF, and we can now consider using these models as a starting point for further structure refinement. PDFGUI contains a useful feature for fitting dPDFs from a crystal structure. For example, load one of the Tb phosphate structures into the fit tree, and select the phase. Then, under the `Configure tab`, you will see a field: `Included Pairs`. Hover the mouse over the field, and the input options will be displayed. You can type `Tb-all` to calculate the Tb dPDF for the given structure. Try inputting some parameters and calculate the dPDF for the two structures; you will see how the resulting PDF changes when different pairs are included.

Once you can simulate dPDFs comfortably, this functionality can also be used to refine a model to the experimental dPDF. By fitting over a short range, you can test whether the local environment of Tb atoms in `H-Zr_2-1_PP_withTb` is similar to the local environment of Tb atoms in either of the structures provided. Try it out. Use Q_{damp} and Q_{broad} given above. In our analysis, we chose to fit the dPDFs over ranges of 1.0–12.0 and 4.0–12.0 Å, and refined a, b, c, the scale factor, the sp_{diameter}, and a single U_{iso} for all atoms. What do the results tell you about the terbium coordination?

12.7   Problems

1. Define turbostratic disorder. What types of materials might be subject to turbostratic disorder, and in what ways can this kind of order manifest?

2. What might be some difficulties in modelling a turbostratically disordered material? In the current example, what assumptions allowed us to use a single-layer model for the structure? What are some ways that this model could be improved?

3. On careful inspection of the PDFs from `H-Zr_2-1` and `H-Zr_1-0`, what differences can be observed, and what do these differences tell you about the ordering of the structure between the two samples?

4. In Chapter 10, the implications of different scenarios for the validity of dPDF analyses were discussed. In that chapter, the scattering from a second, incoherently mixed phase, i.e. the substrate, was subtracted. List some differences with the current case, in terms of what is being subtracted, and what assumptions are being made in doing so.

5. In comparison to the different Tb phosphate structures provided, which structure gave the best match? What does this mean in terms of the structuring of Tb ions between the layers. Use structure visualization software to inspect the structures and look for similarities between them and the α-ZrP layers.

12.8   Solution

12.8.1   Define turbostratic disorder. What types of materials might be subject to turbostratic disorder, and in which ways can this kind of order manifest?

Turbostratic means "unordered layers". This term indicates that three-dimensional correlations in a material are lost by random orientation of layers about the layer normal; it was initially suggested by Biscoe and Warren (1942) to describe the structure of non-ordered carbons. In fact, various extents of turbostratic disorder can occur in a wide variety of materials with layered motifs, which manifest through random layer shifts or rotations, parallel to the adjacent layers.

The loss of full three-dimensional periodicity leads to a loss of *hkl*-type reflections in the diffraction pattern. Despite this *orientational* disorder, *hk*0-type reflections still appear as broad, triangular peaks due to the two-dimensional crystallinity within the layer. Finally, because the periodic stacking of the layer is still maintained, the 00*l*-type reflections remain as sharp crystalline Bragg peaks. It is more difficult to predict how the PDF will look for turbostratic materials, and this depends on how distinctly the neighbouring layers interact. If interacting in very distinct ways, the PDF may still appear as for a crystalline material but with new peaks representing the atom-pair correlations resulting from different neighbour-layer associations. As the possibilities increase, only sharp peaks associated with intralayer atom-pairs may remain visible, with the interlayer component appearing like a damped sine wave representing the density modulation between layers, as in the current study.

12.8.2   What might be some difficulties in modelling a turbostratically disordered material? In the current example, what assumptions allowed us to use a single 2D layer model for the structure? What are some ways that this model could be improved?

Turbostratic materials can pose some challenges when analysed with small box models. The reason for this is the statistical nature of the disorder, both in terms of the different neighbour-layer associations which can manifest and in terms of the frequency and order in which they occur. In simple cases, the neighbour-layer associations may be specifically known through interlayer chemical constraints. However, even for just a few associations, the number of possibilities in the relative arrangement over several layers can quickly become huge. This often requires the simulation of diffraction effects from thousands of models constructed with different probabilities of associations. With the PDF, an approximation can be made where models representing the possibilities of different layer associations just within a certain refinement range (e.g. over just a few layers) can be constructed to represent all possible atom-pair associations and summed to approximate the faulting density (Yang *et al.*, 2013).

In the current example, we assumed that the turbostratic disorder was sufficient for interlayer atom-pair correlations to be neglected. In other words, we assumed that the associations between any given neighbouring layers are completely random, so that on average, all interlayer atom-pair associations are averaged out. Thus all sharp peaks in the PDF come from intralayer atom-pair distances. Realistically, this assumption is probably not exactly correct, further hinted at by the not-extremely-good R_w values and features in the residual curve, but it was sufficient to identify the inorganic α-type layer as the primary building block of the structure and get an estimate of the domain size.

Another consideration involved cutting the 002 peak in the diffraction pattern out prior to Fourier transformation to remove the broad interlayer density modulations in the PDF. In fact, for the H-Zr_2-1 sample, these modulations were very weak in the PDF obtained from the full diffraction pattern, indicating that the periodicity of the interlayer spacing was actually quite weak. For H-Zr_1-0, the effect of the 002 peak was stronger and non-negligible. An improved model may incorporate the stacking of multiple layers such that their orientations can be refined.

12.8.3 On careful inspection of the PDFs from H-Zr_2-1 and H-Zr_1-0, what differences can be observed, and what do these differences tell you about the ordering of the structure between the two samples?

The most prominent difference is the presence of an obvious long-wavelength oscillation with a wavelength of approximately 10 Å in H-Zr_1-0. This comes from the sharp 002 diffraction peak and is an indication of the periodic arrangement of the stacked layers. For H-Zr_2-1, the Bragg peak is still present, but much broader, so we know that there are some stacking correlations; however, they are not visually obvious in the PDF pattern. These oscillations can only be seen when we extract the density distribution function from the first peak alone. Both of these factors contribute to the assessment that the number of layers stacked is few and that the stacking is less ordered, indicating either a larger distribution of stacking distances or the presence of other types of disordering of the individual layers with respect to their neighbours. We also note that the oscillation for the 2:1 sample has a longer wavelength (and Bragg reflection at lower angle) than the 1:0, which indicates that the average stacking distances are shorter for the more strongly interacting/ordered layers.

Some other subtle differences are present. These differences can be observed as increased relative intensities in the 1:0 sample PDF at $r = 10.2$, 11.75, 15.0, and 17.9 Å, for instance. These peaks are not well described by our single-layer structure model and therefore may represent interlayer atom-pair distances which do have a coherent structural relationship. The fact that the intensities at these distances is not zero in the 2:1 sample, may further indicate that some weak atomic-level ordering between the first few neighbouring layers does occur, even in the most turbostratically disordered samples. You should have obtained a poorer fit of the single-layer model (the H-Zr_1-0 sample), which is further evidence for increased interlayer ordering.

12.8.4 In Chapter 10, the implications of different scenarios on the validity of dPDF analysis were discussed. In that chapter, the scattering from a second, incoherently mixed phase, i.e. the substrate, was subtracted. List some differences of the current case, in terms of what is being subtracted and what assumptions are being made in doing so.

In the former case, the scattering components are incoherent phases, so the coherent scattering only comes from intraphase atom-pairs. In other words, the components are only of type A_i-A_i from phase A or B_i-B_i from phase B. If we subtract phase

B, all that remain are A_i-A_i correlations. In the current case, the subtraction is of a host structure from a coherent component of a single phase, meaning that we have three types of components, A_i-A_i from the host phase, B_i-B_i from the guest components, and also A_i-B_i from guest–host correlations. In this case, subtraction of A leaves us with both B_i-B_i and A_i-B_i components. Unlike the previous case, this situation can no longer be analysed separately from the total structure.

12.8.5 **In comparison to the different Tb phosphate structure provided, which structures gave the best match? What does this mean in terms of the structuring of Tb ions between the layers. Use structure visualization software to inspect the structures and look for similarities between them and the α-ZrP layers.**

The best result was found to be the hypothetical Tb phosphate Scheelite-type structure. Looking at the Scheelite structure, we can see that Tb^{3+} ions prefer an oxygen coordination of eight, and thus we can assume that Tb^{3+} ions will also prefer to maximize oxygen coordination in H-Zr. This explains why Tb^{3+} ions would prefer to sit in vacant sites between the layers, which have a similar local environment in the Scheelite-type structure with maximal coordination by oxygen dangling from phosphate groups, and possibly some coordination of neighbouring Tb^{3+} ions.

13

Magnetic PDF

Benjamin A. Frandsen

13.1 Introduction and overview

Neutrons scatter not only off atomic nuclei but also off any magnetic moments. Thus, neutron PDF data collected on magnetic materials will contain both the atomic PDF (sometimes referred to as the nuclear PDF in the neutron world and the function we have been focusing on throughout this book) as well as the magnetic PDF (mPDF). Sometimes, the mPDF signal is so small compared to the atomic PDF that it is not easily observable. However, at other times it is clearly observable and can be used to learn about the local magnetic correlations in the material, opening up a whole new realm for investigation with PDF techniques. The goal of this chapter is to extract and model the mPDF signal from neutron PDF data collected on MnO, which exhibits antiferromagnetic order below 118 K. We will also model the atomic structure along the way. We will use both PDFGUI and DIFFPY-CMI in this chapter.

13.2 The question

MnO is a well-known antiferromagnetic insulator. The Mn^{2+} ions possess large, localized spins with $S = 5/2$ that are paramagnetic above 118 K but order into the arrangement shown in Figure 13.1 below 118 K (Shull *et al.*, 1951). As seen in the figure, all the spins lie within the (111) planes of the crystal structure. They align parallel to each other within their (111) plane, but the spin directions are reversed between adjacent planes. This gives an overall antiferromagnetic structure. This ordered pattern of spins results in additional Bragg peaks in the neutron scattering pattern that get included in the Fourier transform to generate the experimental PDF. Therefore, the overall experimental PDF pattern contains both the atomic and magnetic PDF.

There are many interesting scientific questions related to MnO, but the simple question we wish to answer here is whether we can observe and model the

Atomic Pair Distribution Function Analysis. Benjamin A. Frandsen, Oxford University Press.
© Simon J. L. Billinge and Kirsten M. Ø. Jensen (2023). DOI: 10.1093/oso/9780198885801.003.0013

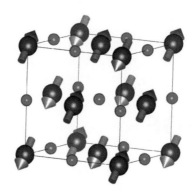

Figure 13.1 Atomic and magnetic structure of MnO. Purple and red spheres are Mn and O atoms, respectively. The arrows represent the Mn^{2+} spins.

magnetic PDF signal contained in experimental PDF data collected on MnO at 15 K (Frandsen and Billinge, 2015).

13.3 The result

From our analysis, we confirmed that the atomic structure can be well described by the published rhombohedral model, which is equivalent to a compression of the cubic structure along the [111] diagonal. More importantly, we confirmed that a strong mPDF signal can be seen in the total PDF data as a large, well-defined signal in the atomic PDF fit residual. The mPDF can be accurately modelled using the published antiferromagnetic structure of MnO. These results, as well as a more detailed analysis of the local atomic and magnetic structure both above and below the antiferromganetic transition at 118 K, are discussed more fully in Frandsen and Billinge (2015) and Frandsen *et al.* (2016a). A theoretical introduction to mPDF analysis is provided in Frandsen *et al.* (2014).

13.4 The experiment

The experiment was performed on a finely ground powder sample of MnO under the conditions shown in Table 13.1. The files you will need for the chapter are listed in Table 13.2.

13.5 What next?

1. Download the files.
2. Get a CIF file from PDFITC to use as a starting point.
3. Create a PDFGUI project, and start a new fit for the atomic structure.

Table 13.1 Experimental details.

Facility	Lujan Center at the Los Alamos Neutron Science Center
Beamline	NPDF
Date collected	October 2013
Sample geometry	Packed powder in standard vanadium sample can
Sample environment	Closed cycle refrigerator, 15 K
Total exposure time	2 hr
Q_{max}	35.0 Å^{-1}
Q_{damp}, Q_{broad} from calibration	0.01426 Å^{-1}, 0.02406 Å^{-1}

Table 13.2 Files to download.

Filename	Note
`npdf_07334.gr`	Experimental PDF data
`MnO_cubic.cif`	CIF file for cubic MnO structure
`mPDFintro.ipynb`	Jupyter notebook containing a tutorial for the diffpy.mpdf software

4. Create the structure for the fit by loading the CIF file, load the data file, and set the fit range to $0.5 \leq r \leq 50.0$ Å.

5. Fix the instrumental parameters to appropriate values, and set the refinable parameters to be variables, including the overall scale factor, the cubic lattice parameter, the rhombohedral angle, ADPs, and δ_1 or δ_2. You can use the space group information to load the appropriate ADPs automatically, or just use isotropic ADPs for each atom type because this is approximately a cubic structure. There are no positional degrees of freedom for the structural model we are using.

6. Perform a calculation with the starting structure to compare with the data and make sure it is at a good enough starting point. Use this to estimate a sensible starting scale factor.

7. Refine your model to obtain the best atomic PDF fit. If all goes well, you should end up with R_w around 30%. Plot the measured PDF, the best-fit PDF, and the difference curve. The magnetic PDF signal will be evident in the difference curve because, by default, the mPDF is not included in the model of the atomic structure.

8. Export the fitted PDF and the fitted structure to use for the mPDF refinement.

9. Create a Jupyter notebook or Python script in which to do the mPDF refinement using DIFFPY.MPDF. At the time of writing, this package is not built into the standard DIFFPY-CMI installation, so you will need to install it separately. Visit https://github.com/FrandsenGroup/diffpy.mpdf for instructions on how to do that.

10. Create a model of the magnetic structure using the refined atomic structure and the published propagation vector of (1/2, 1/2, 1/2) (in the pseudocubic setting) with a spin direction of (1, -1, 0).

11. Create an `MPDFcalculator` and link it to the magnetic structure.

12. Set up a fit recipe that will optimize the agreement between the calculated mPDF pattern and the mPDF data (really just the residual from the atomic PDF fit), with the paramagnetic and ordered scale factors treated as refinable parameters. Inspect the fit, and verify that it quantitatively reproduces the data.

13.6   Wait, what? How do I do that?

13.6.1   Download the files

Follow the instructions in Section 3.6.1.

13.6.2   Get a CIF file from PDFITC to use as a starting point

Using STRUCTUREMINING in PDFITC is a great way to get CIF files to make a start on your refinement campaign. Please see Section 6.6.2 for more details. For convenience, we also provide the CIF with the download data.

13.6.3   Create a PDFGUI project and start a new fit

Open up PDFGUI, and create a new fit. Refer to the earlier description in Section 3.6.3 if you need a refresher on how to do this in PDFGUI. Give your fit a descriptive name, and save the project for safekeeping. We recommend frequent saves, which gives quicker recovery from crashes.

13.6.4   Load the structure and data files into your fit

Referring to 3.6.4 if necessary, load the structure from your CIF file. Verify that the phase information has been populated correctly; you can make any necessary changes directly within PDFGUI. Add the dataset in the same way as outlined in Section 3.6.3, and verify that it has read the correct scatterer type (neutrons) and Q_{max} value in the *Configure* panel. Change the fit range to be from 1.5 to 50 Å.

13.6.5 Fix the instrumental parameters to appropriate values, and set the refinable parameters

In the dataset configuration tab, set the appropriate values for Q_{damp} and Q_{broad} (listed in Table 13.1).

Now we will set all of the parameters that will be refined as variables during our modelling of the atomic PDF. First is the overall scale factor, which you can set in the dataset configuration tab or the structure configuration tab (see Section 5.6.4). Now we want to add the structural parameters as variables. We will take the low-temperature rhombohedral compression of the MnO lattice into account in the modelling. Click on the phase in your fit tree, and go to the `Constraints` menu. Set one variable for the a, b, and c lattice parameters (they should all be identical in the structural model we are using), and set another variable for the three angles of the unit cell (again, they should all be identical, but they may vary from 90° as we want to refine the rhombohedral compression). Make sure to use a different number after the @ for each independent variable. Set one variable for all the diagonal ADPs (i.e. U_{11}, U_{22}, and U_{33}) of the Mn atoms and another variable for the diagonal ADPs of the O atoms. Choose either δ_1 or δ_2 to refine.

If you want to see how we set up the refinement, you can peek at our `.ddp` file distributed along with the data, but we highly recommend doing it all by yourself!

13.6.6 Perform a starting calculation

Just as we did in Section 3.6.7, we will begin by calculating what the PDF from the starting structure will look like. Select your fit, go to the `Calculations` tab at the top of the window, and select `New calculation`. The calculation will appear in the fit tree. Select the calculation, and press the blue gear icon to do the calculation. To view the calculation, select the calculation from the fit tree and then press the `Plot` icon on the top task bar. You can also plot the data by selecting the data from the fit tree and pressing the same plot icon. Visually compare the calculation and data to make sure they look approximately the same. You can also compare the magnitudes to get a reasonable starting value for the scale factor, which may help the initial fit converge better. In this case, a scale factor of 1 appears to be a good starting point.

13.6.7 Refine the model

You are now ready to refine the model. If necessary, refer back to Section 3.6.7 for a reminder about how to run refinements in PDFGUI. The refinement process works best when you take a sequential approach: first refine the parameters that have the biggest effect on the calculated PDF, then add more and more variables until you have the best possible fit. You can fix and unfix variables using the checkboxes in the fit pane, and you can copy the refined variable values to the initial parameter values for each new cycle of refinement in the way described in Section 3.6.7. If you need a hint for the refinement sequence, see the Solution section. Inspect the fit by clicking on the view icon. To view the fit residual clearly, you may need to offset it by about -600, which you can set in the Plot Control pane. Our fit and fit residual are plotted in Figure 13.2, where you can clearly see a large residual, representing the magnetic PDF.

13.6.8 Export the fitted PDF and structure

We now want to export the fitted PDF and the refined structure so they can be used for the mPDF refinement. To export the fitted PDF, click on the data file in the fit tree and then select `Data` → `Export Fit PDF` from the menu at the top of the screen. Choose a descriptive name for the `.fgr` file and save it in the directory where you are doing your analysis. To export the refined structure, click on the structure in the fit tree and then select `Phases` → `Export Fit Structure` from the menu at the top of the screen. Choose a descriptive name for the `.stru` file, and save it in the same directory as the `.fgr` file.

13.6.9 Create a Jupyter notebook or Python script

We will use the DIFFPY.MPDF package to perform the mPDF refinement, so for this we will need to create a Jupyter notebook or Python script and import the appropriate packages. Note that you will need DIFFPY-CMI installed, including DIFFPY.MPDF (see https://github.com/FrandsenGroup/diffpy.mpdf for installation instructions), to proceed with this exercise. We will assume a sufficient level of familiarity with Python and Jupyter to create your own scripts and notebooks. Since this chapter is not intended to be a user manual for DIFFPY.MPDF, we recommend that you first work through the DIFFPY.MPDF tutorial Jupyter notebook included in the downloads for this chapter. The tutorial will give you a good idea of how to create magnetic structures and calculate the corresponding mPDF pattern. After working through the tutorial, you should be able to proceed with the mPDF analysis of MnO.

13.6.10 Create the magnetic structure model

Create a `MagSpecies` object using the refined structural model from your atomic PDF fit. Then set the propagation vector (the `kvecs` attribute) to (0.5, 0.5, 0.5) and the spin direction (the `basisvecs` attribute) to (1, -1, 0), or any other direction perpendicular to the (1, 1, 1) direction. This is the accepted magnetic structure, which can be found in the literature. Be sure to set the `rmaxatoms` attribute to 50 Å (or whatever the maximum r value of your atomic PDF fit is). Then create the `MagStructure` object, load in the `MagSpecies` object, and generate the atom and spin arrays.

13.6.11 Create an MPDFcalculator and link it to the magnetic structure

Create an `MPDFcalculator` object and pass the `MagStructure` object to it. Set the r_{min} and r_{max} attributes of the calculator to the minimum and maximum r values of your atomic PDF fit. Calculate and plot the "non-deconvoluted" mPDF pattern, since this is what is generated experimentally by the standard PDF data reduction protocols. This calculates the magnetic PDF without deconvoluting the effects of the magnetic form factor. Load in the mPDF data (in this case the fit residual from the atomic PDF fit) using the `getDiffData` function with your exported fitted PDF file. Compare the calculated mPDF pattern to the data, and adjust the paramagnetic scale factor and ordered scale factor (`paraScale` and `ordScale` attributes of the `MPDFcalculator` object, respectively) to get a reasonable agreement by eye. The paramagnetic scale factor is related to the self scattering from an isolated spin and is therefore present even in the paramagnetic phase. It results in a peak at very low r, even lower r than the nearest neighbor spin–spin distance. The ordered scale factor reflects the strength of the spin–spin correlations.

13.6.12 Perform the mPDF fit

Load the mPDF data into a Profile object. Define a function that uses the `MPDFcalculator` to calculate the mPDF pattern when given values for the paramagnetic scale factor and the ordered scale factor. Create a `FitContribution` for the mPDF, load the `Profile` object into it, and register your function that calculates the mPDF. Then set this function as the equation for this `FitContribution`. Finally, create a `FitRecipe`, add the `FitContribution`, and add the paramagnetic and ordered scale factors as variables. The fit can now be performed in the usual DIFFPY-CMI way using standard least-squares optimization routines. Overlay the best-fit mPDF and the data to verify a quantitatively accurate fit.

13.7 Problems

1. How do the positions of the features in the mPDF pattern relate to Mn–Mn atomic pair separation distances? You will notice that the mPDF signal is much broader in real-space than the atomic PDF signal. This is because the magnetic scattering is strongly suppressed with increasing Q due to the magnetic form factor, and the standard PDF data reduction protocols do not attempt to correct for this effect by normalizing by the magnetic form factor. For this reason, we call the experimental mPDF the "non-deconvoluted" mPDF.

2. What happens if you fix the paramagnetic scale factor to 0 and do not refine it? Which part of the fit suffers from this? Does this have any relevance for any spin–spin correlations in MnO?

3. The magnitude of the ordered magnetic moment can be determined by comparing it to the atomic PDF scale factor. For spin-only magnetic moments, the ordered moment in Bohr magnetons is given by $m = 2\sqrt{\frac{B\langle b\rangle^2}{An_s}}S_{\text{nominal}}$, where A and B are the atomic PDF scale factor and magnetic PDF ordered scale factor, respectively, $\langle b\rangle$ is the average nuclear scattering length of the atoms in the material, n_s is the fraction of atoms in the system that are magnetic (just Mn in this case, so $1/2$), and S_{nominal} is the magnitude of the spin vectors used in your magnetic structure for the mPDF refinement. Calculate m from your refinements (you will have to look up the appropriate nuclear scattering lengths). How does it compare to the ideal value of 5 μ_B for perfectly ordered $S = 5/2$ spins? It should be slightly reduced from the maximum value due to fluctuations.

4. Give your magnetic model a random starting direction for the sublattice spin vector (i.e. the `basisvecs` attribute of the `MagSpecies` object), and refine the two polar angles θ and ϕ as free parameters. Does the refined spin direction lie in the expected plane perpendicular to the [111] direction?

13.8   Solution

In Figure 13.2(a), we display the best atomic PDF fit for MnO, which we obtained after refining the scale factor, lattice parameters, ADPs, and δ_1. The refinements clearly show the rhombohedral compression, as the peak positions cannot be well

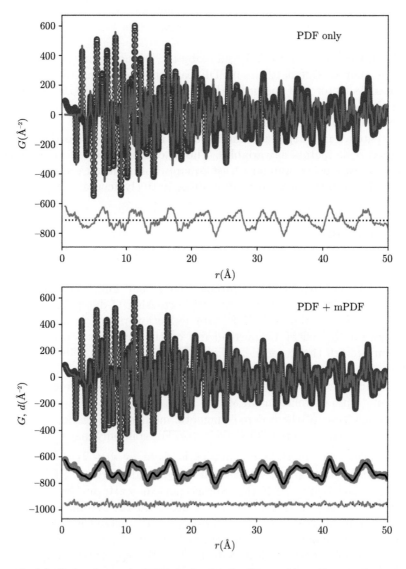

Figure 13.2 (a) Optimal atomic PDF fit for MnO obtained by refining the scale factor, lattice parameters, ADPs, and δ_1 over a range from 0.5 Å to 50 Å. The green curve is the fit residual, offset vertically for clarity. The fit residual contains the experimental mPDF signal, which was not included in the atomic PDF model. (b) Optimal fit with the mPDF contribution included. The top set of curves is the atomic fit shown in panel (a), while the grey and black curves below represent the atomic PDF fit residual (i.e. the mPDF data) and the calculated mPDF pattern, respectively. The green curve on the bottom is the overall fit residual when both the atomic PDF and mPDF contributions are included.

fitted if the rhombohedral angle is not included as a variable. Note that the fit residual for the atomic PDF is extremely large ($R_w = 30.86\%$) and highly structured, which proves that it is not just random noise. Instead, the fit residual contains the experimental mPDF data, since the atomic structural model did not include any magnetism. Therefore, we used the atomic PDF fit residual as the experimental data against which to refine the magnetic model. We fixed the sublattice spin direction to be (1, -1, 0) and refined only the paramagnetic scale factor and ordered scale factor for the magnetic model. The refined parameters are provided in Table 13.3. Together, the calculated atomic and magnetic PDF describe the experimentally measured PDF with a high level of accuracy ($R_w = 7.31\%$).

We note that the positions of the features in the mPDF pattern roughly correspond to Mn–Mn atomic pair separation distances, but the significant broadening of the mPDF signal compared to the atomic PDF signal means that most peaks or valleys in the mPDF pattern actually contain contributions from many pairs of spins. This is especially true as r increases. It is also worth pointing out that the refined value of the ordered scale factor corresponds to an ordered moment of 4.63 μ_B, which is slightly lower than the full 5 μ_B expected for $S = 5/2$ spins due to magnetic fluctuations and other effects. Finally, the refined spin direction is very robust in the case of MnO. If we start the refinement with a random spin direction, the fit still converges to a spin direction within the expected plane perpendicular to the (1,1,1) direction. This shows that mPDF analysis can be highly sensitive to spin orientations within the crystal lattice.

Philosophically, it would be more satisfying to do a simultaneous co-refinement of both the atomic and magnetic structure against the total PDF data. This can be done fairly straightforwardly using DIFFPY-CMI, and we recommend doing this when possible. Examples of corefinements can be found in the tutorial resources available at https://github.com/FrandsenGroup/diffpy.mpdf. However, in practice there is often no significant difference between the results obtained from a genuine co-refinement and a sequential refinement such as the one performed in this chapter.

Table 13.3 Refined parameters from the atomic and magnetic PDF modelling of MnO at 15 K.

a (Å)	4.4307
$\alpha = \beta = \gamma$ (°)	90.6096
U_{Mn} (Å^2)	0.00228
U_O (Å^2)	0.00312
δ_1 (Å^2)	0.6942
Atomic PDF scale factor	0.9518
Atomic PDF R_w	0.3086
DIFFPY.MPDF paramagnetic scale factor	2.9053
DIFFPY.MPDF ordered scale factor	1.2073
Atomic + magnetic PDF R_w	0.0731

This is because the atomic PDF and mPDF signals are quite distinct from one another, so there is typically little to no correlation between the parameters describing the atomic structure and those describing the magnetic structure.

That concludes this chapter. Congratulations! You have now completed your first mPDF refinement. We hope you find many future opportunities to use mPDF analysis!

14

Tips and tricks: PDF measurements

Maxwell W. Terban

14.1 Introduction and overview

While this book is focused on how to extract science from a PDF once it has been acquired, here we give a brief overview of the considerations in data acquisition needed to obtain high-quality PDFs. Comprehensive literature on these topics is available (Egami and Billinge, 2012; Billinge, 2019; Terban and Billinge, 2022). However, we hope that a less formal discussion will help highlight some practical aspects of the process, without belabouring the theory, and help you to optimize your experiment or better communicate your needs to a beamline scientist.

14.2 Basic overview: What are total scattering data?

In order to perform PDF analysis, a total scattering measurement must first be performed. This is like a typical powder diffraction measurement, but with additional considerations. Total scattering measurements are generally performed by transmitting a high-energy beam of collimated radiation through a sample and measuring the intensity of the scattered radiation signal. Total scattering means that both Bragg and diffuse scattering intensities (the intensities underneath and between the Bragg peaks) must be collected with good statistics. The data collection must cover as much of reciprocal-space as possible, since a requirement of total scattering is that the data must be acquired over a wide enough Q-range to ensure that the atom-pair distances in the PDF can be suitably resolved. In general, the Q-range of a scattering measurement can be tuned to probe density distributions in the material at varying length scales. For example, PDF analysis requires wide-angle scattering (WAS) with a high Q_{max} value (generally in the range 15–40 Å^{-1}) to resolve distinct atom-pair distances, while small-angle scattering (SAS) may give information on nanometre-scale heterogeneities, for example on the distribution of molecular motifs (Mou *et al.*, 2015) or particles in nanoparticle assemblies (Liu *et al.*, 2020).

Atomic Pair Distribution Function Analysis. Maxwell W. Terban, Oxford University Press.
© Simon J. L. Billinge and Kirsten M. Ø. Jensen (2023). DOI: 10.1093/oso/9780198885801.003.0014

14.3   What type of radiation should I use?

Total scattering data may be collected from beams of x-rays, neutrons, or electrons. Each source of radiation has different strengths and weaknesses, and these are discussed in Chapter 2. As well as access to a particular source of radiation, other considerations include the visibility of atoms of interest in your sample (for example, light atoms in materials containing heavy atoms may be more visible in a neutron measurement than an x-ray one), the real-space resolution you require (synchrotron x-rays and neutrons tend to give higher real-space resolution than lab x-rays or electrons), the amount of sample you have (electrons < x-rays < neutrons), the beam sensitivity of your samples (neutrons are less damaging than x-rays which are less damaging than electrons), and the penetrability you require of your radiation, for example, getting into special environments (neutrons > synchrotron x-rays > lab x-rays > electrons). Finally, we note that, at the time of writing, PDFs obtained from x-rays and neutrons are more quantitatively reliable than from electrons due to uncorrected dynamical scattering effects in ePDF measurements coming from the significant multiple-scattering effects. However, highly valuable quantitative information may be obtained from ePDFs despite the dynamical scattering, especially if steps are taken to minimize the dynamical effects by making samples thinner, focussing on less well-ordered materials, and precessing the incident beam (Hoque *et al.*, 2019). Recent reviews on the topic further discuss the details of these measurements and analysis (Gorelik *et al.*, 2019; ao Batista Souza Junior *et al.*, 2021). Indeed, modern approaches are allowing quantitative crystallography to be carried out with electrons (Gemmi *et al.*, 2019), and PDF studies will also benefit from these.

The rest of the chapter will give tips and tricks on setting up total scattering experiments using synchrotron x-rays, to help guide novice users in these decisions. Whether using synchrotron x-rays or neutrons, instrument scientists will be most knowledgeable about their particular instrument and its capabilities.

14.4   Detectors

As an x-ray beam passes through a sample, the radiation from all scattering events sums to form a propagating wavefront which is emitted from the sample. The goal in a powder total scattering experiment is to collect the intensities of this wavefront as a function of the magnitude of the scattering angle (the angle between the incident beam and the scattered beam). This can be accomplished in various ways and has historically been done with a wide array of different detector technologies, which we will not cover here (Egami and Billinge, 2012; Billinge, 2019). In principle, a perfect detector would be a hollow sphere surrounding the sample, the interior consisting of a radiation-sensitive material with infinite spatial resolution for detecting scattered photons. Such a detector would cover 4π steradians of solid angle and is sometimes called a "4π detector". The solid-angle coverage of any real detector will be less than 4π steradians, but increasing the solid-angle coverage is often an aspirational goal, especially in measurements where production of the probe particles is expensive, as is the case for neutrons, for example. In practice, detector geometries are somewhat limited by the respective technologies used, but as time goes on they get better and better.

14.4.1 Area detectors

Most x-ray total scattering measurements are done in the so-called rapid acquisition PDF (*RA-PDF*) mode in which a large-area 2D detector is used to collect a two-dimensional cross section of reciprocal-space (Chupas *et al.*, 2003). This was introduced in Chapter 2, and Figure 2.1 shows a sketch of an *RA-PDF* experiment. The demonstration in 2003 that large-area 2D detectors could capture diffraction patterns suitable for PDF analysis in a single shot (one detector position) effectively revolutionized the application of PDF methods, reducing experiment times from more than 12 hours to less than a second, a more than four orders of magnitude increase in data throughput. At the time of writing, common area detectors used at synchrotron beamlines for PDF analysis include amorphous silicon flat plate detectors and hybrid photon counting devices. Disadvantages of the 2D detectors are their finite physical size, meaning that measurements can only be obtained over a given angular range depending on the sample-to-detector distance. High $Q_{\max}$ values ensuring good r-resolution in the PDF can be achieved with high-energy x-rays and short sample-to-detector distances, but this leads to a low Q-resolution. This does not present a significant problem if the scientific question under study involves studies of the very short-range structure (< 10 nm) of the material. Indeed, currently much of the scientific impact of PDF studies comes from such cases. In principle, area detectors could be scanned in space to increase the angular range. However, the engineering and data analysis logistics of this are quite complicated and not often pursued.

The *flatness* of the detectors generates some systematic effects. It results in an angle-dependent intensity correction due to a higher detection efficiency at higher angles, as well as a change in high-angle peak positions (Marlton *et al.*, 2019). This situation can occur if the x-rays penetrate all the way through the thin photo-active layer, which is common in high-energy synchrotron x-ray experiments. These effects may be corrected in the data integration and correction process. Other corrections and considerations for optimal measurement with area detectors have been reviewed elsewhere (Skinner *et al.*, 2012a).

Area detectors can be positioned in different ways with respect to the incident beam path in order to vary the obtainable $Q_{\max}$, Q-resolution, and count rate of the measurement. The simplest placement is with the incident beam at the detector centre (Figure 14.1a), wich gives the best statistics as full Debye–Scherrer cones are collected. Increasing the sample-to-detector distance gives higher Q-resolution but decreases the usable $Q_{\max}$, thereby reducing the PDF r-resolution. This may be a good trade-off if you are particularly interested in sample morphology (size and shape of particles) but don't need super-high real-space resolution to separate nearby PDF peaks. The detector can also be set to be perpendicular to the incident

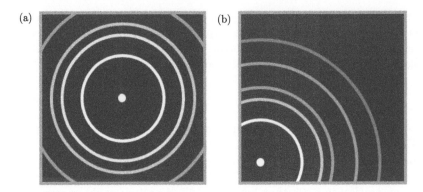

Figure 14.1 Sketch of detector images with the incoming beam in the centre of the detector (a) and in the corner of the detector (b).

beam but offset laterally, so that the beam points toward the corner of the detector (Figure 14.1(b)), which can help increase the Q_{max}, at the cost of count-rate. More exotic oblique geometries have also been suggested (Burns *et al.*, 2023).

14.4.2   Point and linear detectors

Point detectors are an alternative to area detectors. They must be used in accordance with angle-dependent scanning and give the best, and consistant, Q-resolution. The downside is that such scans are exceptionally slow. The throughput can be increased by using diffractometers with multiple point detectors, such as multi-analyser crystal setups, for example at beamline ID22 at the ESRF or P02.1 at DESY (Schökel *et al.*, 2021). It is possible to get very high Q-resolution at these instruments and therefore high-quality data for careful measurements of sample morphology. However, the experiments are still quite slow, with no advantage gained for analysis of the local structure signal in the low-r region, which is often the primary rationale for doing PDF measurements. The slower speed is especially an issue for *in situ* or *operando* experiments. Linear detectors can provide a balance between measurement time and resolution effects. However, they are still quite slow compared to large area detectors. One solution is to panel many linear detectors into arrays, which has been implemented, for instance, with the Mythen detectors developed at the Swiss Light Source (Bergamaschi *et al.*, 2010) or the Arc-Detector at the XPDF beamline at Diamond Light Source (Gimenez *et al.*, 2019), though these are not the most commonly used detector technologies. Most common at the time of writing are still large-area amorphous silicon detectors, but hybrid photon-counting detectors such as the Pilatus detector (Broennimann *et al.*, 2006) are rapidly growing in adoption.

Some general considerations on the detection technology include the dynamic range and the detection efficiency. These can be important depending on the energy used in your experiment. Additional corrections, such as for intrinsic electronic noise (often called dark current or dark counts), non-uniform spatial response (often called flat-field corrections), or non-linearities in the detection response are also needed (Kern, 2019). These corrections are generally known and applied at the beamline, and you can ask your beamline scientist for details.

14.5 Sample geometries

Technically, all standard sample geometries used for powder diffraction are possible for total scattering with varying considerations and corrections applied (Egami and Billinge, 2012). Since area detectors in a transmission (forward-scattering) geometry are used in many cases for xPDF and ePDF measurements, we eschew further discussion of measurement geometry here and instead give a brief overview of the types of sample geometries and environments that are common for total scattering studies.

14.5.1 Capillaries

For most total scattering measurements of powders, suspensions, liquids, gels, etc., x-ray total scattering data are measured in a capillary transmission geometry. This is not the most well-matched sample geometry for the *RA-PDF* geometry – "hockey-puck" or pill-shaped geometries make more sense. However, the practical advantages of using capillaries almost invariably outweigh the geometry consideration. These advantages include that samples can be easily prepared and sealed prior to the measurements, so open materials do not necessarily need to be shipped to or handled at the beamline. Capillaries are also easily placed in and removed from a goniometer for quick sample exchange. In fact, many x-ray total scattering beamlines have high-throughput sample changers which are optimized for running many capillary samples automatically without the need for moving in and out of the experiment hutch between measurements. Capillaries are also easily spun to minimize effects from sample texture or coarseness due to the presence of large particles. They also tend to be easy to use, with various sample environments for heating or cooling of the sample. In cases where powder-type samples cannot be prepared, pieces, chunks, or bits of material can be sandwiched between Kapton film or tape and held perpendicular to the incident beam.

14.5.2 Thin films

As we saw in Chapter 4, x-ray total scattering data from thin films can also be measured using transmission geometry (Jensen *et al.*, 2015). A possible setup is sketched in Figure 14.2. The high flux at synchrotron light sources allows data with sufficient scattering statistics to be collected on both film and substrate, such that the substrate scattering can be directly subtracted. To obtain the best data quality, the ratio of thin film to substrate scattering should be maximized, principally by using an amorphous substrate but also by selecting weakly scattering substrate materials, such as disordered polymer or glass, and to make them as thin as is feasibly possible. With that said, if the experiment requires a thicker substrate, the high fluxes at synchrotrons and advanced data reduction methods allow surprisingly small signals to be extracted (Terban *et al.*, 2015). Amorphous substrates are favoured for simplifying the background subtraction, though it is possible to subtract a polycrystalline substrate signal as long as there is a consistent powder average. Subtraction of background scattering from single-crystal substrates is much more difficult as it is likely to result in imperfectly subtracted Bragg peaks in the signal and should be avoided where possible. Ideally, the background scattering should be measured from the same substrate as used for the deposited sample, or a highly similar one, to decrease the likelihood of any deviation in substrate thickness, texture, structure, and so on.

In general, significantly longer data collection times are needed for thin film samples compared to powders in capillaries due to the very small amount of material yielding the interesting scattering signal. This goes for all types of samples where the material of interest gives rise to only a small fraction of the scattering signal.

Recently, grazing incidence geometry has been used to obtain PDFs from very thin films (Dippel *et al.*, 2019). Grazing incidence measurements increase the relative signal from a thin film, at the cost of a more specialized and complicated experiment. This geometry also makes spatially resolved measurements on thin films more difficult because of the large footprint of the beam on the sample and the sensitivity of its position on sample angle (Kovyakh *et al.*, 2021). Normal incidence tfPDF is a simple and fast and versatile option if you have sufficient signal from your film. It was recently demonstrated in an electron microscope mapping local structural features with a spatial resolution of less than 10 nm (Rakita *et al.*, 2022).

14.5.3   Containerless methods

A rather exotic, but worth being aware of, option for sample containment is levitation. For example, an acoustic levitator uses sound waves to levitate small pieces or liquid droplets of a material. In this case, the levitated sample removes the need for a physical container, minimizing the background signal which must be separately measured and subtracted (Weber *et al.*, 2009; Weber *et al.*, 2012). In addition, it enables following structural changes or reactions in solution without any liquid–container interface effects (Thi *et al.*, 2015). A schematic representation of an acoustic levitator is given in Figure 14.2. Other levitation methods using, for example, electric fields are also possible, as well as non-static containerless methods, such as measuring streams of liquid or droplets (Skinner *et al.*, 2012b).

14.5.4   Computed tomography and PDF: ctPDF

By combining PDF with computed tomography, it is possible to perform PDF experiments on just about any object, allowing 2D slices or 3D reconstructions of local structure as a function of position in solid objects (Jacques *et al.*, 2013), Figure 14.2.

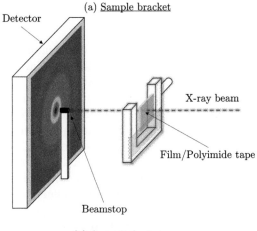

(a) Sample bracket

Detector

X-ray beam

Film/Polyimide tape

Beamstop

(b) Acoustic levitator

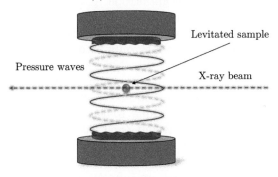

Levitated sample

Pressure waves

X-ray beam

(c) Computed tomography (CT)

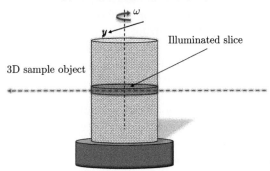

ω

y

Illuminated slice

3D sample object

Figure 14.2 Different sample geometries: (a) thin film transmission sample geometry, (b) acoustic levitation as used for small droplets or beads of material, and (c) computed tomography setup for spatially resolved data. Figure reproduced with permission from Terban and Billinge (2022).

14.6   Samples

14.6.1   Sample preparation

Sample preparation is particularly important in ensuring high-quality data. Obtaining a good *powder average* is critical to quantitative reliability. Ideally, the sample would be orientationally isotropic, meaning that either the atomic structure of the sample is isotropic (most liquids, gels, or amorphous solids) or that the crystallite size is sufficiently small, approx. 1–10 μm. Larger crystallites can be ground to ensure that powders are not coarse or textured, though care needs to be taken to minimize modifications to the sample by the grinding. In the case of capillaries, the sample can be spun about its axis (perpendicular to the beam) to increase the distribution of crystallite orientations sampled by the beam. In order to obtain high-quality data, the packing efficiency in the capillary can be maximized ensuring the most material in the scattering volume for improved statistics. With that said, it is important not to introduce a preferred orientation or texture, which can happen, for example, when highly anisotropic powder grains lie along a shared axis. Sometimes issues such as these are only discovered at the beamtime. For example, non-uniform (textured sample) or spotty rings (too large grain-size) or insufficient material in the capilliary may only be evident when the sample is put in the beam. It is always a good idea to have extra sample, and extra sample holders, on hand at the experiment to try to mitigate these things. However, it is possible to recover from many sample issues after the fact with modern data reduction methods, such as auto-masking routines that can statistically remove pixels with outlier intensities (Wright and Zhou, 2017), see https://xpdacq.github.io/xpdtools/. Also, as discussed earlier in the book, it is possible to study textured samples in real-space (Gong and Billinge, 2018; Harouna-Mayer *et al.*, 2022). Since synchrotron and neutron beamtimes are infrequent and precious, paying careful attention ahead of time to sample preparation, for example having samples that were ground by different amounts or approaches, and also having mitigation strategies in case these things show up during the experiment, is the best approach. It may only become apparent after the experiment whether data from an unground or heavily ground sample are better for answering the scientific question. 2D detector diffraction images collected from various samples in various modalities are shown in Figure 14.3.

Finally the sample or capillary thickness can also be increased or decreased depending on the measurement conditions. Thinner (smaller diameter of capillary) samples will provide better Q resolution but may require longer measurements and increase the chance for texture or spottiness in the diffraction pattern. For lower x-ray beam energies, x-ray absorption can also pose a problem for thick samples made up of high-Z elements. As a rule of thumb, many high energy x-ray experiments are done on 0.5–1 mm diameter capillaries for well-scattering samples, but often 2 mm and even 3 and 4 mm diameter capillaries and tubes designed for nuclear magnetic resonance (NMR) experiments are used for weakly scattering samples. Your beamline scientist is likely to be able to give advise on these matters for your specific samples.

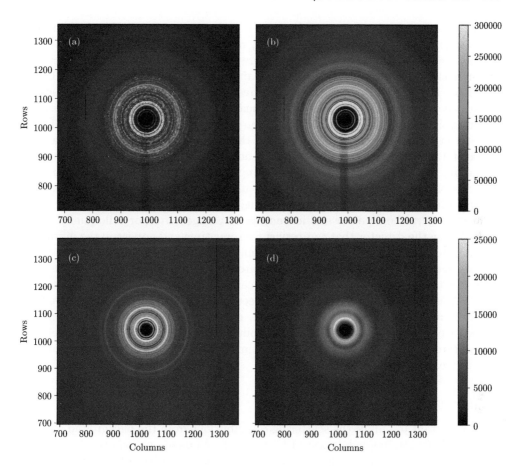

Figure 14.3 2D scattering intensities measured from (a) polycrystalline S_8 without spinning, (b) polycrystalline S_8 with spinning, (c) polycrystalline α-lactose monohydrate, and (d) amorphous spray-dried lactose.

14.6.2 Standards

Apart from collecting data from your samples, you always need to also collect data from a standard sample. Measurements from a standard are used to calibrate the instrumental parameters for data integration and reduction and to determine the instrumental parameters relevant for PDF modelling, i.e. Q_{damp} and Q_{broad}. Good standards for PDF analysis include Ni, Si, CeO_2, or other highly crystalline standard materials from NIST. These measurements should be performed every time the experimental setup is modified, such as changes to the sample position, detector position, or beam size. Standards should ideally be measured in the same diameter capillaries and experimental geometry as the samples of interest to ensure that the profile function for the instrument and sample, and its effects on the resulting

PDF, can be properly accounted for in subsequent data analysis. Even if the setup is consistent, it is good practice to regularly measure a standard to ensure consistent energy stability after any event where there are significant changes in ring current or monochromator temperature, such as after injections or beam dumps. This is particularly required for highly quantitative analyses, but if the science question doesn't depend on a precise independent measurement of lattice parameter, as a last resort, it is possible to auto-calibrate by varying calibration parameters until some feature in the measured PDF appears at the right location (for example, a silica peak at 1.6 Å). However, this is never recommended if it is possible to measure a standard. For some detectors, it may be important to attenuate the incident beam flux by placing attenuators in the beam and measure data from your standard sample for only very short times. A strongly scattering sample with sharp Bragg peaks can burn the detector and leave trapped charge states, or colloquially "ghost images", on the detector that can persist for some time and appear in the diffraction pattern from a subsequent sample measurement. This is much more of an issue with weakly scattering samples, and if your materials are weakly scattering, take great care not to burn the detector with your calibration sample, or any other sample for that matter. Ghost images on the detector can be reduced over time by doing multiple readouts with the shutter closed, or spread out over Q by changing the detector position (Skinner *et al.*, 2012a), but depending on how badly the detector was saturated, they can last a very long time and quite possibly longer than your beamtime. Take great care with this.

14.6.3   Background measurements

It is important to reduce the scattering contributions from the background as much as possible to obtain good data for PDF analysis. As mentioned above, the capillary or film material used for sample support would ideally consist of a thin-walled, non-structured, and weakly scattering material such as polyimide (Kapton) tube or tape, or borosilicate glass. For high temperatures, there are different more temperature-resistant silicas and quartz capillaries.

The scattering signal from the background (e.g. an empty capillary, a clean film subtrate, or an empty sample environment) must be measured with good statistics. Since the data obtained from background measurements will be subtracted from the data from your sample when obtaining the PDF, you ideally need at least as good background data as sample data to avoid introducing noise into your PDF. If the background is much lower in intensity than the signal, this is less of an issue, but it is highly amplified if the signal of interest is a small proportion of the entire signal, as in a tfPDF experiment. During your beamtime, make sure to collect data from all the background signals you may need in your data analysis. For example, if you want to do a dPDF analysis where you subtract the contributions from one of the components from the PDF, you should collect high-quality data for each of the components at the right conditions. Ideally, for each component, measurements should be made of the background scattering contributions from the sample container, air scattering (no sample holder or sample), and an empty sample environment if you are using one. It is sometimes necessary to make linear combinations of these to get a proper subtraction. For example, even though the empty sample holder signal includes the air scattering and should give a good subtraction of both, if, for example, there was a variation of the amount of capillary silica in the beam between the sample and background measurements, you may need a different scaling to remove both the silica and the air scattering. This is handled by taking a linear combination of these and optimizing the fractions of each. The functionality of subtracting linear combinations of different backgrounds is present in the latest version of PDFGETX3. It is always better to come home with too many backgrounds than too few!

The incident beam (and the transmitted and diffracted beams) also scatters from air along the beam path. To minimize this, the incident beam upstream of the sample is typically encased in a collimator, a tube made of absorbing material that absorbs most of the scattered radiation, and it is beneficial to extend this tube to a finely collimated point as close to the sample position as possible. This is typically done by the instrument scientists at a beamline. However, very-high-energy x-rays are difficult to completely block, and it can sometimes happen that shadows of pieces of equipment and beamline components are observed in the measurement. Placement of shielding (for example lead sheets) at different locations can minimize this and is typically incorporated into the setup of the hutch by the beamline scientists. However, things can change when new experiments are set up, or improved shielding may be necessary for highly sensitive measurements. Then, a trial and error process of improving the shielding may be necessary. It can take some time but is well worth it. Especially if you are measuring weakly scattering samples, it is worth spending time at the beginning of the experiment to minimize these effects rather than find out at the end that they have ruined your data. They are most evident in scattering patterns from sample backgrounds and in air scattering measurements.

Air scattering occurs from the incident beam after the sample in high-energy x-ray experiments. To reduce this, it is possible to move a beam-absorbing beamstop closer to the sample and further from the detector. This comes at the price of losing low-angle information but could be a good tradeoff if there is no interesting scattering signal in the low-Q region. For larger sample-to-detector distances, it may be useful to use a secondary flight-path tube consisting of a large-diameter tube or cone that contains a weakly scattering gas like helium, or is completely evacuated. In this design the tube has a large enough diameter that it contains not only the transmitted incident beam but also all the scattered beams as well. That is, it would need to have a diameter which is the size of the detector at the position of the detector. This is similar to what is done for small angle scattering beamlines but can be warranted for longer sample-to-detector distances even in medium- or wide-angle experiments. You can discuss the possibilities for your measurements with the beamline scientist when preparing for beamtime.

14.7 Sample environments

Various sample environments and apparatuses are available at different beamlines for the study of material under diverse and highly specialized conditions. These tend to be listed at beamline websites, but it is always best to reach out and talk directly to the beamline scientists who have a better idea of what will be the best environment for your particular needs. Heating and cooling devices are standard at most total scattering beamlines. For instance, cryocooling systems are common in ranges from 80–500 K, which utilize streams of cold nitrogen gas pulled from liquid nitrogen dewars. These also now exist with the cooling being done by cold helium gas instead of nitrogen. These have base temperatures as low as 4 K but can be difficult to use at these low temperatures due to icing and other factors. All the cryostream coolers only cool a small spot, so there are large temperature gradients across your sample. It is therefore important to take care to line up the cryostream and x-ray beams carefully. Improved thermal stability at ultracold temperatures can be obtained by putting the sample in a vacuum and placing heat shields around it. Such devices tend to be called cryostats rather than cryostreams and will give a more uniform temperature across your sample. However, the experiments can be much slower as it takes longer for the sample to reach equilibrium at each temperature.

More specialized setups have been developed for combining additional methods with total scattering, for example coupled acoustic levitation and Raman scattering (Thi *et al.*, 2015), or combined PDF and IR spectroscopy (Beyer *et al.*, 2014). Many other potential setups for different environmental conditions exist, such as microwave reactors, furnaces or high pressure cells, as well as combined methods, either already at beamlines or designed, built, and brought by experimentalists. A recent innovation is the gradient furnace, in which variable-pitch windings along the length of a cylindrical sample allow the temperature along the sample to vary over a wide range (for example 200–700 °C) (O'Nolan *et al.*, 2020). It is then possible to carry out a temperature-dependent measurement very rapidly simply by moving the sample across the beam. You will probably have to get any self-built equipment safety tested before being allowed to use it at the beamline.

14.7.1   *In situ* and *operando* experiments

PDF is an excellent technique for studying structural changes during a process such as a chemical synthesis, a catalytic reaction, or battery cycling. Several research groups have developed specialized setups where such changes can be studied, and some are available for general users at synchrotron beamlines. For example, a thin capillary reactor has been used to study nanoparticle formation in solvothermal synthesis (Becker *et al.*, 2010), and capillary gas flow cells have been used to study catalytic reactions (Chupas *et al.*, 2008; Becker *et al.*, 2010). A sputtering chamber has been developed for *in situ* grazing incidence PDF of thin film growth (Roels-gaard *et al.*, 2019), and several setups have been developed for *operando* studies of battery materials (Chapman, 2016).

The data quality you can obtain from many *in situ* and *operando* experiments are often far from what you get from a powder sample closely packed in a thin capillary. The experimental conditions needed for your *in situ* measurements do not always allow you to follow the guidelines for sample optimization that we described above, and when designing setups for *in situ* experiments, you often have to compromise between data quality, experimental conditions, and time resolution in your measurements. Importantly, it is crucial to optimize the amount of interesting material in the beam. for example, if you are studying reactions in solution, you may aim for a high concentration of the material in question so that you do not just measure a scattering signal from the solvent. In studies of material synthesis in solution, this means that the precursor solution should be more concentrated, if feasible for your study, than that which might be used for synthesis in your home laboratory. In the same way, if studying the reactions taking place during battery cycling, you must think about how you can optimize the signal from the structures you are interested in while still having a battery that works. Like any other experiment, discuss the possibilities with the beamline scientists and seek advice in the literature and from others in the field.

15

More PDF tips and tricks

Kirsten M. Ø. Jensen and Simon J. L. Billinge

15.1 Introduction

In this final chapter, we give a few more tips and tricks that you may be able to use when you start your own PDF analysis. We also describe some new developments taking place in PDF-land that could help shape the future of material structural analysis.

15.2 PXRD or PDF, Q-space or r-space analysis?

At this point we hope it is pretty clear that PDF is incredibly useful for structural analysis of a range of different materials, from local structure in bulk materials to disordered nanomaterials and small clusters. However, before you start planning your own experiments, it is important to consider in general which technique(s) and modelling strategies will be the most useful for your particular project.

Total scattering and PDF analysis are not the first characterization techniques you will use after synthesizing or getting a new sample. Many chemistry or materials science labs have access to a conventional powder diffractometer where you can collect PXRD data, and this is usually the go-to method to get the most preliminary information about whether your synthesis went as planned, if your sample is crystalline, nanostructured, or amorphous, and if you have the expected crystalline phases in your sample. It is a good idea to establish this before you start planning any PDF experiments. If it turns out to be crystalline, it is also highly recommended that you consider treating your PXRD data with Rietveld refinement so that you can have a quantitative analysis of the crystalline structures in your sample. Rietveld refinement is highly complementary to the PDF modelling described in this book, and having thoroughly modelled your PXRD data, you will have a much better starting point for PDF analysis and for deciding if PDF analysis is the way to go for your structure characterization. Some good resources for learning more about Rietveld analysis are, for example, Dinnebier and Billinge (2008) and Dinnebier *et al.* (2019).

Atomic Pair Distribution Function Analysis. Kirsten M. Ø. Jensen and Simon J. L. Billinge, Oxford University Press.
© Simon J. L. Billinge and Kirsten M. Ø. Jensen (2023). DOI: 10.1093/oso/9780198885801.003.0015

After an initial analysis of your sample, you can then carefully consider what characterization strategy you will use to answer all your scientific questions. In some cases, the best method is thorough analysis of high quality powder diffraction data. If, for example, you are looking for accurate unit cell parameters from a crystalline sample, medium or high resolution PXRD is generally the way to go. As mentioned in Chapter 5, identification of minority crystalline phases can readily be done in reciprocal-space due to the clear separation of Bragg peaks. Nevertheless, as we have seen in this book, PDF analysis of bulk crystalline materials can reveal local structures different from the average structure found from a Rietveld refinement, so it might be useful to do both!

In Chapters 3 and 8, we used PDF to characterize nanoparticle size along with atomic structure. PDF is well-suited for size analysis of small nanoparticles, usually below 10 nm. The upper limit for reliable size determination depends on the instrumental damping, which comes from the Q-space resolution of the measurement. Because we often value flux and Q-range over Q-space resolution, standard *RA-PDF* measurements are not optimized for finding particle morphologies for larger samples. The larger the damping arising from the instrument, the lower the limit for reliable crystallite size determination. If you are interested in analysing crystallite sizes between 15 and 100 nm using scattering data, you may therefore want to do a high-Q-resolution powder diffaction experiment with Rietveld refinement, where crystallite size and strain can be extracted from the Bragg peak shapes (Dinnebier and Billinge, 2008). This is sometimes referred to in the literature as size-strain analysis. You may also search for papers on full-profile fitting of powder diffraction data. Reciprocal-space analysis of crystallite size can also be favourable if you have a highly anisotropic crystallite shape (rods, platelets or similar), as you can directly relate the apparent crystallite size to specific crystallographic directions.

At most of the synchrotron beamlines optimized for PDF experiments, you can do complementary PXRD experiments with high Q-resolution by simply moving the detector further away from the sample. You can therefore have your cake and eat it, by measuring the scattering signal from the sample at two detector positions (near and far) to obtain data optimized for PDF and for Q-space fitting. Some beamlines (for example PDF (28ID-1) at NSLS-II) are moving in the direction of making this more routine and easy by having two detectors set up at different differences and cutting a slot out of the near detector to allow a quadrant of the scattering signal to reach the far detector. When you are planning your PDF experiments, consider if this may be useful for your scientific questions and discuss the possibilities with the beamline scientist.

15.3 Model-free analysis of PDF

One of the major strengths of PDF is that structural analysis is done in r-space, allowing for an intuitive, model-free approach. As we have seen throughout the book, this is often the first step in any PDF study. By inspecting your PDFs and comparing your data to calculated PDFs from known structures, you are able to get a lot of information on your sample by identifying which pairs may give rise to the peaks you see. Below we suggest some more strategies and tools for getting the most information from your model-free analysis.

15.3.1 PDFmorph

We often want to compare two or more PDFs to see whether there has been any meaningful structure change from one measurement to the other, for example a phase transition or a structural distortion. This may be done by plotting the two PDFs on top of each other and plotting the difference curve below the PDF curves. This is easily done in XPDFSUITE, in a Python session, or using any other plotting program. You will often see lots of features in the difference curve, which makes it look as if the sample has undergone a structural modification. However, if the two datasets were collected at different temperatures, as is often the case, some (or all) of those features come from non-interesting factors, such as peak shifts due to thermal expansion, peak broadening due to increased thermal motion, or simply that the two PDFs are not on exactly the same scale for some reason. To tackle this issue we have developed a program PDFMORPH. This program will take one PDF file as a reference and then take another PDF file that you want to compare to the first, and it will apply different simple transformations to the second PDF in a regression loop ("morphing") to try and get the second PDF to come into as good agreement as possible with the first PDF. This is like fitting, but it is model-independent, and it is designed simply to mimic the effects we mention above. It will allow the second PDF to be "stretched" along the r-axis to mimic thermal expansion, "smeared" by broadening the PDF to mimic increased thermal motion, and "scaled" to correct any scaling errors. Any signal in the difference curve that is left over after morphing is probably due to an interesting structural modification. There are also a few other morphing transforms in PDFMORPH.

PDFMORPH can be installed from conda-forge, and the code is in the diffpy.org directory at GitHub.

15.3.2 Model-independent peak extraction

Sometimes we want to extract the distances between atoms in our solid, but we do not have a good structural model (if we have a good structural model, all of the interatomic distances can be obtained from our model, directly in PDFGUI, as shown in Section 12.6.11). To address this issue we have developed a program SRMISE (Granlund *et al.*, 2015), which is also available at diffpy.org and through conda. This program uses the Akaike information criterion to do a sparse extraction of peaks from the PDF, where the PDF is decomposed into the smallest number of (Gaussian) peaks that are justified based on statistical grounds. A drawback of this program is that it requires some human involvement in estimating the PDF baseline. A new, fully automatic peak extraction program is currently under development and will be available at PDFITC and through conda.

15.3.3 Principal component analysis and other multivariate analysis algorithms

When doing *in situ* or *operando* experiments, you often come home from synchrotron experiments with gigabytes or terabytes of data. If you have structure models for of all the components in your changing samples, you can can set up sequential modelling in PDFGUI or DIFFPY-CMI and extract all structural information from the datasets. However, quite often, new structures will show up that you may not know or that you cannot fully model. Principal component analysis (PCA) or non-negative matrix factorization (NMF), or other related methods, can in this case help you separate contributions in the PDF signal and understand the changes taking place during the experiments or between your datasets. We will not go into any detail here, but the point in all these methods is to describe a series of datasets with a number of fundamental components. These components can either be the phases present in the changing samples or changing features in the PDF. Multivariate analyses of PDFs have been used in the data analysis for a range of different kinds of experiments, for example for identifying changes in phase distributions during chemical reactions and in understanding how phase transformations take place (Chapman *et al.*, 2015; Olds *et al.*, 2017). We have found that the NMF decomposition algorithm is particularly useful for PDF studies (Liu *et al.*, 2021). An app, NMFMAPPING, (Thatcher *et al.*, 2022) has been uploaded to PDFITC and is an easy way to do an NMF decomposition on your data.

15.4 More options for PDF modelling

15.4.1 PDFGUI or DIFFPY-CMI?

Throughout the book, we have shown how to analyse PDF data in both PDFGUI and DIFFPY-CMI. As discussed already in Chapter 1, PDFGUI is much easier to use, and we recommend everyone doing PDF analysis to start out with their projects in PDFGUI, which is intuitive and where it is much easier to get an overview of how changes in the model affect your fit. However, DIFFPY-CMI gives many more possibilities for modelling, for example for discrete structures like clusters and layers which cannot formally be modelled in PDFgui, although some tricks can be done (Chen *et al.*, 2023). DIFFPY-CMI has many more functions than described here, and you can build very advanced models. The strength of DIFFPY-CMI is its flexibility. You can, for example, fit particle size distributions and particle shapes other than the spherical model available in PDFGUI (Gamez *et al.*, 2017) (which could, for example improve the fit of the Pt nanoparticles in Chapter 3). If your sample contains both discrete clusters and crystalline particles, you can create a model with both structures (Juelsholt *et al.*, 2019; Jensen *et al.*, 2012), or you can add any mathematical function to your fits. You can fit PDFs from molecular materials and take into account intra- and intermolecular interactions (Prill *et al.*, 2015; Prill *et al.*, 2016). Basically, the possibilities in DIFFPY-CMI are endless! With the scripts distributed with this book, we hope to make it easier to get started with DIFFPY-CMI. More scripts are available in the CMI exchange on GitHub (https://github.com/diffpy/cmi_exchange), where we encourage users to share their own scripts with the rest of the community. Remember also to use the diffpy-users Google Group mailing list (https://groups.google.com/g/diffpy-users), where you can ask (and answer) questions of the community. We encourage you to request to join the community.

15.4.2   Complex modelling

Material characterization rarely relies on one technique alone; most often, you will combine multiple approaches – for example, scattering techniques with electron microscopy and spectroscopy methods to get a better understanding of your sample. When modelling data to obtain a structural model, it can be useful to take this a step further and analyse datasets from different experimental techniques in combination, i.e. using a single model to fit to data from different probes. This is referred to as "complex modelling" (Juhás *et al.*, 2015) or "multi-modal analysis" and is useful when one single technique cannot provide enough information to yield a robust or unique structure model on all length scales. For example, small angle scattering (SAS) is very sensitive to nanoparticle size, size distribution, and shape, while PDF is very sensitive to atomic structure. By combining data from SAS and PDF, and modelling the two datasets with one single model, we can get more robust results. This of course assumes that the measurements are done under similar conditions and that the same type of model can be used when fitting data, i.e. it must be parameterized in the same way. This approach has been used to model the structure of CdS quantum dots (Farrow *et al.*, 2014) and can be implemented in DIFFPY-CMI.

15.5   Automated PDF modelling

When doing your PDF modelling, you may want to try out many models to see which fits best to your data. This could be the case for crystal structures or cluster structures, where you probably want to test out many different but related models. Instead of doing this manually, this process can be automated by "mining" structures from databases or building them from known building blocks. For example, this has made it possible to automatically determine the atomic structure in the cores of metallic nanoparticles (Banerjee *et al.*, 2020) or to identify structural motifs in disordered metal oxides (Lindahl Christiansen *et al.*, 2020b; Anker *et al.*, 2022). By using Python and DIFFPY-CMI, you can create scripts that take structures from databases of CIFs or automatically built files with xyz-coordinates of atoms and fit PDFs from all of them to your data. Instead of manually testing a handful of models, this means that you can quickly test thousands of models and extract information on which models fit best and what the best fitting models have in common. In terms of statistics, this places a greater emphasis on model selection over parameter estimation, which is what is being done in PDFGUI, where the goal is to estimate values for a relatively large number of parameters from a single model. Many such automated modelling strategies are under development. In this modality the application of machine learning can be quite beneficial to speed up the search, and automated modelling is very likely to play a much larger role in PDF analysis in the future.

15.5.1 Other approaches and programs for total scattering analysis

This book has focused on the use of small-box modelling for PDF analysis. We have shown how to use PDFGUI and DIFFPY-CMI, but other software packages exist where similar analysis can be done, such as TOPAS (Coelho, 2018), which is currently widely used for reciprocal-space Rietveld refinement but can also be used for PDF analysis. The program DISCUS (Proffen and Neder, 1997) can also be used for small-box modelling, and multilevel analysis can be introduced, where things such as ensembles of particles can be built for refinement of stacking fault densities or particle shapes and dispersions.

A mentioned in Chapter 1, many other strategies for PDF analysis exist. For example, in "big-box" modelling methods, the box of atoms used in the model (which in small-box modelling is often the crystallographic unit cell) is enlarged, so that it contains tens of thousands of atoms which can be moved independently (i.e. without crystallographic symmetry constraints) in the modelling to obtain the best fit to the data. This can be done using the reverse Monte Carlo algorithm, and by setting constraints for the atom movement based on, for example, chemically relevant interatomic distances and motifs, structure models for, for example, disordered or fully amorphous solids or liquids can be obtained. Such kinds of analysis can be done using RMCprofile (Tucker *et al.*, 2007), EPSR (Soper, 1996), or Fullrmc (Bachir, 2016), among other programs.

We also note that total scattering data can be analysed in Q-space, where the Debye equation is used to calculate scattering patterns from nanoparticles and even nanoparticle ensembles, whilst structural parameters are refined. DIFFPY-CMI has a Debye calculator that computes the Debye function before Fourier transforming it to get the PDF. For fitting larger models that include defects, surface effects, and morphologies, the DEBUSSY program suite (Cervellino *et al.*, 2015) is a powerful option.

15.6 Final words

The examples shown in this book have been selected both to get you through the beginning steps of learning PDF analysis and at the same time to show the enormous versatility of applications of PDF. This goes for the types of materials that can be studied and for the types of experiments that can be done. We have focused very much on examples from our own work, but the field of PDF has grown enormously. There are many powerful developments in the methodology made by other groups. We do not wish to diminish all these contributions, and we encourage readers to explore and enjoy the full breadth of work that is out there. The goal of this book was to give people a scaffold for taking the first steps in the method rather than providing a complete overview of the field, and we had sufficient examples for this from our own work, which of course is more familiar to us. We have tried to briefly mention other software and capabilities, and it is our hope that once you get started with PDF you will have enough knowledge from this book to try out these other programs as well.

There are also, at this point, literally thousands of PDF studies of a wide range of materials, from food science to conservation science, through pharmaceuticals, all the way to energy materials, catalysts, quantum materials, and exotic superconductors. Now that you know how to do your PDF analysis, we encourage you to dive into the PDF literature, learn more about the uses of PDF, get started developing your own PDF studies, and make your own contributions to that literature! Good luck on your PDF journey.

Appendix A: Python

A.1 Introduction

Over the years, for our PDF software we have used many software languages, from FORTRAN (Chivers and Sleightholme, 2018), through the once-cutting-edge YORICK (Munro and Dubois, 1995), to now almost exclusively Python (Van Rossum and Drake, 2009) and C++ (Stroustrup, 1996). Python is rapidly becoming the language *du jour* for much of scientific data analysis, and it is well worth spending some time to learn the basics. Much of the software we use in the book has been written in Python, and therefore you need a Python distribution installed on your computer to run them. You can also write your own Python scripts, which is very useful for any scientist. There are many online resources for learning Python, and Google can help you find them. Some courses that people seem to like include Real Python (https://realpython.com/), Learn Python the Hard Way (https://learnpythonthehardway.org/), Google's own Python class (https://developers.google.com/edu/python/), and Automate the Boring Stuff (automatetheboringstuff.com). If you want to take your scientific programming to a higher level, transitioning from just writing hacked-together scripts to releasing real open-source software packages to the community, we also recommend *Effective Computation in Physics* by Scopatz and Huff (http://shop.oreilly.com/product/0636920033424.do).

A.2 Installing Python programs

It turns out that distributing software to users is one of the hardest jobs in the world (who knew?). The reason is basically that, as a developer, everything works fine on your local machine when you build the software package, but you are going to send it to someone, and they will run it on their computer, and who knows what has happened to that computer since that person got it, and even before? Do they have a different Python distribution than you? We have all had that annoying error message "myprogram can't find file dothat.cmpgd" or something. One way around this is to send a complete executable that contains everything it needs to run in a binary bundle. This is what happens when you install a program using a Windows installer executable or an `.msi` file on the Mac. This approach makes it difficult to maintain and update software, especially with the limited human resources of a research group, and we use a different intermediate solution. This is to use a package manager and to use "virtual environments" to build up on your computer the exact environment you need for your application to run.

Virtual environments are self-contained environments on your computer where correct versions of all required software are all installed and ready to run a particular application. If you want to run another application that has conflicting needs, you can create a separate environment for that one, and you can have both applications running on your computer, at the same time, by running each one in its own environment. Virtual environments are cool. For example, you can have a "Python 3" environment where all your Python 3 code will run and a "Python 2" environment where all your Python 2 code will run. When you want to run a Python 2 program, you "activate" the Python 2 environment and run the program, and likewise for the Python 3 environment.

Whilst we can protect you from much of this by making executables and have you run code by double-clicking icons on your desktop, it is highly beneficial for your research career if you can spend the time to learn at least the basics of installing and running software in virtual environments in this way as it opens up many more opportunities to work with powerful software packages that are only available this way. This is the case, for example, if you want to write your own Python scripts but make use of libraries such as numpy or scipy. This chapter gives some basics for understanding the rationale behind the approach and the basics for you to learn how to become independent, so you are answering questions on the diffpy-users Google Group rather than asking them and ultimately, if you want, contributing to the programs themselves with bug fixes and feature requests!

A.3 The terminal and the command prompt

Many of you may not have worked with a command-line interface (CLI), sometimes called a terminal, so we provide some basic introduction here. You can open a terminal on Windows by typing `cmd` into the Windows search box. The command line is a place where you can type commands, and if the operating system understands them it will do things. Most importantly, you can run programs from the command line. For example, to run Python (assuming you have Python installed), open a terminal and at the command line type `python`. Your computer will run the Python program. Nothing will happen in this case because you told the operating system to run the Python program, but you didn't give the Python program anything to do. If your terminal gets stuck running a program, you can kill the program by typing `CTL-C` on Windows (it will be something like `CMD-C` on Mac). If instead you type `python --version`, you are asking the operating system to run Python and you are asking Python which version of Python it is. Python will then check which version it is running and let you know by printing it to the terminal. Most program CLIs have built -in help that you can access with `-h` or `--help` (note the pattern, which is common in CLIs: one-dash followed by one letter or two dashes followed by a word). Try typing `python -h` and see what happens. We usually don't run Python this way, so the help function is long and confusing. Don't worry, you aren't expected to read and understand it, but just knowing that at a terminal you can nearly always type `<program name> -h` and you will be given some help on how to use `<program name>` is already empowering. By the way, whenever we use the angle-brackets syntax, as with <program name>, we mean that you replace the angle brackets with something else described inside. For example, in this case we could replace <program name> with python but also with whoami (which is a program that tells you who you are, or more precisely, who the computer user is, currently).

It is beyond the scope of this chapter to describe all the things that you can do at the command prompt, and there are many resources online and in books for learning the basics (including *Effective Computation in Physics* by Scopatz and Huff). In the Billinge group we prefer the Unix versions of the command prompt, which comes native on Linux and Mac but can be installed on Windows by installing, for example, GIT BASH (https://gitforwindows.org).

We like using the Conda package manager. With Conda, all the widely used open-source Python packages are kept in a library on the internet and can be installed and updated (and uninstalled) using simple commands on the command line. In this case we also use the Conda virtual environments, and this is what we will describe in this chapter. There are two choices to get this set up on your computer. Either install ANACONDA (www.anaconda.com) or the lighter weight MINICONDA (https://docs.conda.io/en/latest/miniconda.html). We recommend MINICONDA. Anaconda comes with some tools such as a graphical user interface (GUI) for handling the packages, but it also comes with a lot of "bloat", which is software that Anaconda thinks you will find useful and which it installs in your base environment. We prefer to install Miniconda and then to build up all the other bits and pieces you need by installing them only when they are needed. It keeps things cleaner and can avoid problems with inadvertent clashing versions of software in the future.

It is well worth spending some time reading about and getting to know a bit about virtual environments in general and Conda environments in particular. The basic steps to get started using Conda are:

- If you have not already done so, install Miniconda (https://docs.conda. io/en/latest/miniconda.html) (or Anaconda (www.anaconda.com)) on your computer.
- Create your first Conda environment by opening a terminal and typing conda create --name=<my_environment_name> <options>, e.g. conda create --name=mypy3 python=3. Using this command, you create a new environment called "mypy3" where you can run Python 3 programs.
- Now activate your environment for this terminal session, conda activate mypy3. Once you have done this, all commands you type in this terminal will be run in the mypy3 environment. If you also want to run a different program that needs Python 2, open another terminal, activate the mypy2 environment over there, and all commands you type in that terminal will run programs using Python 2.
- Your first terminal session is still in the mypy3 environment. However, as we just created the environment, it is still pretty empty; there is not much

there except basic Python 3, so let's install some stuff: `conda install numpy, scipy, matplotlib`.

- Now you have Python 3 versions of numpy, scipy, and matplotlib installed. You can go ahead and install anything else you want in there, e.g. DIFFPY-CMI (you will have to follow the instructions at diffpy.org to get the right syntax). DIFFPY-CMI is in a different "channel" on Conda, its own diffpy channel, that you have to ask Conda to look in by typing `--channel diffpy` as part of the command.
- Whenever you want to play with some Python 3 code, just
  - open a new terminal
  - activate your Python 3 virtual environment: `conda activate mypy3`
  - work work work.
- For example, to set up an environment to run PDFGUI:
  - `conda create --name=pdfgui python=3`
  - `conda activate pdfgui`
  - Now, we have to tell Conda where to find PDFGUI. It is in our diffpy conda channel. This command adds the diffpy channel to your configuration so Conda will always look in there if it can't find the program elsewhere: `conda config --add channels diffpy`
  - `conda install diffpy.pdfgui`
  - And you are good to go. You don't need to install numpy or scipy because Conda is clever enough to figure out everything that PDFGUI needs to run (its dependencies) and to install everything it needs, as well as installing the software itself.
- For example, to run PDFGUI after it is set up:
  - open a new terminal
  - `conda activate pdfgui`
  - `pdfgui`

This is just an example for installing a software package. At the time of writing we are updating versions of PDFGUI and also how they are installed, so please make sure to follow the instructions at diffpy.org if these don't work.

A couple of words about 'channels' in Conda. If you don't specify a channel, Conda will only check its own central repository of code (called the default channel). The good news with this is that the code is all trusted and tightly controlled. The bad news is that the code is all tightly controlled! If you make something useful and want to share it, you will have to work really hard to get Anaconda to include it in the main channel. But you can create your own channel, and anyone can download your software if they specify your channel. The good news about that is that you can get all kinds of software from all kinds of developers through their channels. The bad news about that is that you can get all kinds of software from all kinds of developers through their channels, and you have no idea if it is trusted or not. Maybe they are installing a programmatic back-channel to your computer that will siphon all your precious research and sell it to criminals. So when you specifiy a channel that is not the Conda main channel, make sure you trust the source! There is an intermediate channel called conda-forge which is in between (--channel conda-forge). Anyone can seek to get their software onto the conda-forge channel, but there is a community of package maintainers and trusted people. It is by no means 100% secure, but things won't stay around on conda-forge if they are found to be problematic. Also, conda-forge maintains a huge dependency graph of all the packages to make sure that the right version of every dependency is installed. Long story short, if it is on conda-forge, try to install it from conda-forge. You can do this by putting conda-forge as the first channel in your list of channels in the .condarc file in your home directory (you can usually access your home directory by typing cd ~, but it is the directory you are placed in by default when you open a new terminal). When you type conda config <something>, Conda will actually update this config file for you.

```
cd ~            \# to get to your home directory
ls .condarc     \# to list all files with this name. There should be
                just one
less .condarc   \# to see what is in the file
edit .condarc   \# to edit it
```

For example, my .condarc file looks like this:

```
channels:
   - conda-forge
   - diffpy
   - defaults
restore_free_channel: true
anaconda_upload: false
```

Most of the group's software packages are now being released through the conda-forge package repository channel on Conda. You access them by typing at the command line conda install -c conda-forge ⟨ package_name ⟩

A.4 Python IDEs and Jupyter notebooks

A Python script is basically just a text file with a .py ending, and you can use any text editor to create or edit your Python scripts. Python scripts can be run from a terminal by typing python <filename>.py. However, there are lots of "integrated development environments" (IDEs) out there for Python with more or fewer features that help the scripting and programming workflow.

A popular one when you are doing more exploratory work is JUPYTER (http://jupyter.org/). This IDE runs in your web browser (though in general the code is all running locally) and has nice graphics and integrated tools for writing notes. It is kind of half-way house between a full-blown Python IDE and an electronic notebook.

When you move from the exploratory (try this, try that) phase, and towards the exploitative (let's write some reproducible code that really gets the most out of this data) phase, and when you want to be packaging and sharing your code and scripts with others, we recommend that you move away from JUPYTER. There are many full-featured professional-grade IDEs, and you should pick the one you love; one that we currently use a lot is PYCHARM (https://www.jetbrains.com/pycharm). If you are a student or academic, this is free, but it has a commercial version so is very full featured and well maintained. However, it is so full featured that it can sometimes make your computer run slowly. Some lighter-weight but still good ones are Sublime or "Visual Studio Code".

If this is confusing, don't worry; you don't need any of it to run PDFGUI!

Appendix B: Data processing and integration

B.1 Introduction

Here we present a few tips for data reduction from 2D powder diffraction images. This is not meant to be an exhaustive review or a comprehensive step by step approach to data reduction. It is assumed that new users will have help from instrument scientists in performing data reduction, but in the spirit of the book, we hope that providing some additional considerations can be helpful to newcomers.

There are many programs for visualizing and integrating 2D powder diffraction images to 1D powder diffraction patterns. For visualizing and manipulating 2D images, IMAGEJ (https://imagej.nih.gov/ij/index.html) is a popular, powerful, and full-featured tool. Images can also be processed in Python and plotted using MATPLOTLIB.

Integrating 2D powder patterns to 1D is a more specialized activity, and there are a number of programs that have been developed for this, including FIT2D (Hammersley, 2016), SRXPLANAR, which is part of xPDFSUITE (Yang *et al.*, 2015), and GSAS2 (Toby and Von Dreele, 2013). At this point, the most highly developed and most widely used is PYFAI (Kieffer *et al.*, 2015), a fast, open-source, Python-based integration program. GUIs are available for PYFAI, including DIOPTAS (Prescher and Prakapenka, 2015) and a GUI from the PYFAI developers (Kieffer *et al.*, 2020). This chapter introduces some useful terminal commands and GUI functions in PYFAI.

B.2 Image averaging

One philosophy for data collection is, rather than taking one long detector exposure to get sufficient statistics, to take many shorter exposures and sum or average them together. This results in more images to handle but can be useful for verifying sample stability or to avoid losing data if, for example, the beam dumps during the measurement. In this modality, you will have a series of images, typically in `.tiff` (sometimes tiff files will be written with a `.tif` (single "f") extension; make sure to check this in your case) or `.edf` format, which must be combined. With PYFAI, there is an image averaging function which can be used with a single command. It can be used by opening a command terminal and navigating to a directory containing the series of images, and typing

```
pyFAI-average *tiff
```

The new image will be saved in the same directory, with "mean" appended to the filename.

We note that a number of corrections are usually applied to the images before further processing, e.g. dark current correction by subtracting an image taken with the shutter closed or a flat-field correction applied to account for inhomogeneities in the detector response. These corrections are typically applied during data collection, although they will depend on the type of detector technology and operating conditions. More information can be found in, for instance, Lee *et al.* (2008) or Skinner *et al.* (2012a), and it can be useful to discuss these details, or any questions about data format, with the beamline scientist.

B.3 Calibration

The calibration serves to determine the exact geometry of the experimental setup so that images on the detector (in detector coordinates) can be converted to intensities in diffraction space. The main quantity to determine is the effective sample-to-detector distance, but any angular offsets of the detector away from being perfectly perpendicular to the incident beam are also important. Calibration is done by measuring diffraction from a calibrant material and then adjusting the parameters of the detector geometry to give the best fit to the calibrant diffraction pattern. The detector calibration optimizes over all the known d-spacings the calibrant material to determine the distance and non-orthogonality (rotation and tilt) of the detector. For example, calibration of the setup can be initiated using PyFAI by typing

```
pyFAI-calib2
```

in the terminal. The command will open a GUI within which one may load in the desired calibrant diffraction image, set up details of the detector, such as the number and size of the pixels in the detector, define the calibrant, and also generate a mask. Points on any number of rings can be selected and indexed, following which the program will automatically run the refinement of the detector geometry correction parameters. Once finished, a `.poni` file can be saved to the directory, containing the calibrated geometry parameters. These parameters will be used to integrate all other images that were collected using the same sample/detector setup. A list of all options within the command line can be obtained by typing

```
pyFAI-calib2 --help
```

or by visiting the PyFAI documentation website (https://pyfai.readthedocs.io/en/master/).

B.4 Masking and integration

Prior to integration, the image must also be masked to remove invalid pixels and other problematic parts of the image from the integration. Masked pixels will simply not be considered in the integration. If invalid pixels and edges are not masked, the integration will result in a 1D powder pattern with significant errors in the integrated intensities, often appearing as non-physical kinks, spikes, glitches, or other anomalous behaviours.

There are several considerations that might suggest pixels should not be included:

1. Dead pixels: pixels which do not work.
2. Dark pixels: pixels which record a significantly low outlying intensity.
3. Hot pixels: pixels which record a significantly high outlying intensity.
4. Edges: a frame around the edges and any gaps between active detection areas.
5. Shadows: any pixels dimmed by shadows from the beamstop, beamstop holder, sample holder, or downstream equipment, such as a cryostream, photodiodes, or other devices.
6. Non-powder averaged intensities: to the extent possible, the sample should be prepared as an ideal powder. However this is not always possible. Some high-intensity single crystal peaks may remain and should be masked.

Human-drawn geometric masks can be made manually with PYFAI by calling

```
pyFAI-drawmask image.tif
```

and using the interactive plot to create the mask. The general procedure would be to draw polygon masks to remove edges and shadows and then use threshold masking to remove dark and bright pixels.

There are now powerful automasking programs, such as in xpdtools on GitHub (https://xpdacq.github.io/xpdtools/), that use statistical algorithms to automatically remove regions of the image (Wright and Zhou, 2017). After installation, the automasking code can be run by typing

```
image_to_iq -- --help
```

This will result in all possible keyword arguments to be given. The basic requirements are an image or list of images to mask (.tif or .edf format) and a calibration (.poni) file from PYFAI. For example, the full command for automasking will be something like

```
image_to_iq --image-files image.tif --poni-file calibration.poni
```

It is possible to add externally generated or modified masks, as well as controlling the types and parameters of the acceptance criteria. The code will create masks and integrate images automatically and can be easily used to batch integrate a large number of images from an entire experimental session. Examples of image masks can be seen for two different detector types in Figure B.1. The result of the masking

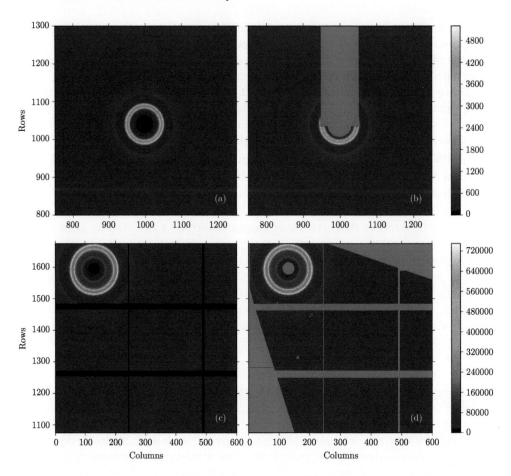

Figure B.1 2D scattering intensities measured from polyamide 6 using an amorphous silicon PerkinElmer detector at the XPD beamline of the National Synchrotron Light Source II, Brookhaven National Laboratory (a) unmasked, (b) masked, and a Dectris PILATUS3 X CdTe 2M at beamline ID31 of the ESRF (c) unmasked, (d) masked.

operations will be saved as arrays that have the same shape as the image arrays but contain 1's or 0's indicating whether a pixel is included or not.

Once the calibration and mask files are generated, the images can be integrated to obtain the 1D powder diffraction patterns. Using the geometry parameters from the standard, every pixel in the image can be indexed to a Q-value. Azimuthal integration around the rings (summed along arcs of constant 2θ or Q) is performed, and the resulting intensities in the final 1D pattern are the averages of all pixels counted in the integration per respective bin. With pyFAI, this is performed using

```
pyFAI-integrate image.tif
```

Another GUI window will open, within which the `.poni` file containing the calibration parameters and the mask file can be loaded. The degree of polarization of the incident beam should be set, typically set to 0.95–0.99 for synchrotron radiation, along with the integration units and number of radial bins for the integrated data. It is best to integrate directly into units of Q (Å^{-1} or nm^{-1}) rather than scattering angle to avoid further systematic signal degradation when resampling data (Yang *et al.*, 2014). More detailed discussions of further corrections and considerations for integrating 2D images can be found in the literature (Hinrichsen, 2007; Hinrichsen *et al.*, 2008; Lee *et al.*, 2008; Skinner *et al.*, 2012a; Yang *et al.*, 2014).

Bibliography

Abeykoon M, Malliakas C D, Juhás P, Božin E S, Kanatzidis M G, and Billinge S J L. Quantitative nanostructure characterization using atomic pair distribution functions obtained from laboratory electron microscopes. Z. Kristallogr., 227(5):248–256, 2012.

Abeykoon M, Hu H, Wu L, Zhu Y, and Billinge S J L. Calibration and data collection protocols for reliable lattice parameter values in electron pair distribution function studies. J. Appl. Crystallogr., 48:244–251, 2015.

Anasori B, Shi C, Moon E J, Xie Y, Voigt C A, Kent P R C, May S J, Billinge S J L, Barsoum M W, and Gogotsi Y. Control of electronic properties of 2D carbides (MXenes) by manipulating their transition metal layers. Nanoscale Horiz., 1(3):227–234, 2016.

Anker A S, Kjær, E T S, Juelsholt, M, Christiansen T L, Skjærvø S L, Jørgensen M R V, Kantor I, Sørensen D R, Billinge S J L, Selvan R, and Jensen K M Ø. Extracting structural motifs from pair distribution function data of nanostructures using explainable machine learning. npj Comput. Mater., 8(213), 2022.

Antonaropoulos G, Vasilakaki M, Trohidou K N, Iannotti V, Ausanio G, Abeykoon M, Bozin E S, and Lappas A. Tailoring defects and nanocrystal transformation for optimal heating power in bimagnetic $Co_yFe_{1-y}O@Co_xFe_{3-x}O_4$ particles. Nanoscale, 14(2):382–401, 2022.

Souza Junior J B, Schleder G R, Bettini, J, Nogueira I C, Fazzio A, and Leite E R. Pair distribution function obtained from electron diffraction: An advanced real-space structural characterization tool. Matter, 4(2):441–460, 2021.

Bachir A. Fullrmc, a rigid body reverse Monte Carlo modeling package enabled with machine learning and artificial intelligence. J. Comput. Chem., 37(12):1102–1111, 2016.

Banerjee S, Liu C-H, Lee J D, Kovyakh A, Grasmik V, Prymak O, Koenigsmann C, Liu H, Wang L, Abeykoon A M M, Wong S S, Epple M, Murray C B, and Billinge S J L. Improved models for metallic nanoparticle cores from atomic pair distribution function (PDF) analysis. J. Phys. Chem. C., 122(51):29498–29506, 2018.

Banerjee S, Liu C-H, Jensen K M Ø, Juhás P, Lee J D, Ackerson C J, Murray C B, and Billinge S J L. Cluster-mining: An approach for determining core structures of metallic nanoparticles from atomic pair distribution function data. Acta Crystallogr. A, 76(1):24–31, 2020.

Barnes A C, Fischer H E, and Salmon P S. Structure of disordered systems and liquid metal alloys. J. Phys. IV, 111:59–96, 2003.

Becker J, Bremholm M, Tyrsted C, Pauw, B, Jensen. K M Ø, Eltzholt J, Christensen M, and Iversen B B. Experimental setup for in situ X-ray SAXS/WAXS/PDF studies of the formation and growth of nanoparticles in near- and supercritical fluids. J. Appl. Crystallogr., 43(4):729–736, 2010.

Beecher A N, Yang X, Palmer J H, LaGrassa A L, Juhas P, Billinge S J L, and Owen J S. Atomic structures and gram scale synthesis of three tetrahedral quantum dots. J. Amer. Chem. Soc., 136(30):10645–10653, 2014.

Benmore C J. A review of high-energy x-ray diffraction from glasses and liquids. ISRN Mater. Sci., 2012:852905, 2012.

Bergamaschi A, Cervellino A, Dinapoli R, Gozzo F, Henrich B, Johnson I, Kraft P, Mozzanica A, Schmitt B, and Shi X. The mython detector for x-ray powder diffraction experiments at the Swiss Light Source. J. Synchrotron Rad., 17: 653–668, 2010.

Bette S, Takayama T, Kitagawa K, Takano R, Takagi H, and Dinnebier R E. Solution of the heavily stacking faulted crystal structure of the honeycomb iridate $H_3LiIr_2O_6$. Dalton Trans., 46:15216–15227, 2017.

Beyer K A, Zhao H, Borkiewicz O J, Newton M A, Chupas P J, and Chapman K W. Simultaneous diffuse reflection infrared spectroscopy and x-ray pair distribution function measurements. J. Appl. Crystallogr., 47:95–101, 2014.

Billinge S J L and Farrow C L. Towards a robust *ad-hoc* data correction approach that yields reliable atomic pair distribution functions from powder diffraction data. J. Phys: Condens. Mat., 25:454202, 2013.

Billinge S J L, and Kanatzidis M G. Beyond crystallography: The study of disorder, nanocrystallinity and crystallographically challenged materials. Chem. Commun., 7:749–760, 2004.

Billinge S J L and Levin I. The problem with determining atomic structure at the nanoscale. Science, 316:561–565, 2007.

Billinge S J L. Local atomic structure and superconducivity of $Nd_{2-x}Ce_xCuO_{4-y}$: A pair distribution function study. Ph.D. Thesis, 1992.

Billinge S J L. Real-space Rietveld: Full profile structure refinement of the atomic pair distribution function. In Billinge S J L, and Thorpe M F, editors, Local Structure from Diffraction, page 137. Springer, New York, 1998.

Billinge S J L. Local structure from total scattering and atomic pair distribution function (pdf) analysis. In Dinnebier R E, and Billinge S J L, editors, Powder Diffraction: Theory and Practice, pages 464–493, Cambridge, UK, 2008. Royal Society of Chemistry.

Billinge S J L. Nanoscale structural order from the atomic pair distribution function (PDF): There's plenty of room in the middle. J. Solid State Chem., 181: 1695–1700, 2008b.

Billinge S J L. The nanostructure problem. Physics, 3:25, 2010.

Billinge S J L. Pair distribution function technique: principles and methods. In Kolb U, David W I F, and Shankland K, editors, *Uniting Electron Crystallography and*

Powder Diffraction, NATO Science for Peace and Security Series B: Physics and Biophysics, pages 179–190, Dordrecht, 2013. Springer Science & Business Media.

Billinge S J L. Nanometre-scale structure from powder diffraction: Total scattering and atomic pair distribution function analysis. In Gilmore C, Kaduk J, and Schenk H, editors, International Tables of Crystallography, volume H, pages 649–672, Chester, UK, Sep 2019. International Union of Crystallography.

Biscoe J and Warren B E. An x-ray study of carbon black. J. Appl. Phys., 13(6): 364–371, 1942.

Božin E S, Knox K R, Juhás P, Hor Y S, Mitchell J F, and Billinge S J L. $Cu(Ir_{1-x}Cr_x)_2S_4$: A model system for studying nanoscale phase coexistence at the metal-insulator transition. Sci. Rep., 4:4081, 2014.

Božin E S, Yin W G, Koch R J, Abeykoon M, Hor Y S, Zheng H, Lei H C, Petrovic C, Mitchell J F, and Billinge S J L. Local orbital degeneracy lifting as a precursor to an orbital-selective peierls transition. Nature Commun., 10:3638, 2019.

Burns N, Rahemtulla A, Annett S, Moreno B, and Kycia S. An inclined detector geometry for improved X-ray total scattering measurements. J. Appl. Cryst., 56:510–518, 2023.

Broennimann C, Eikenberry E F, Henrich B, Horisberger R, Huelsen G, Pohl E, Schmitt B, Schulze-Briese C, Suzuki M, Tomizaki T, Toyokawa H, and Wagner A. The PILATUS 1M detector. J. Synchrotron Radiat., 13(2):120–130, 2006.

Brown I D, and Altermatt D. Bond-valence parameters obtained from a systematic analysis of the Inorganic Crystal Structure Database. Acta Crystallogr. B: Struct. Sci. Cryst. Eng. Mater., 41(4):244–247, 1985.

Cervellino A, and Frison R. Texture corrections for total scattering functions. Acta Crystallogr. A, 76(3):302–317, 2020.

Cervellino A, and Frison R. Texture corrections for total scattering functions. Acta Crystallogr. A, 76(3):302–317, 2020.

Cervellino A, Giannini C, and Guagliardi A. DEBUSSY: A Debye user system for nanocrystalline materials. J. Appl. Crystallogr., 43:1543–1547, 2010.

Cervellino A, Frison R, Bertolotti F, and Guagliardi A. *DEBUSSY 2.0*: the new release of a Debye user system for nanocrystalline and/or disordered materials. J. Appl. Crystallogr., 48(6):2026–2032, 2015.

Chapman K W, Lapidus S H, and Chupas P J. Applications of principal component analysis to pair distribution function data. J. Appl. Crystallogr., 48(6): 1619–1626, 2015.

Chapman K W. Emerging operando and x-ray pair distribution function methods for energy materials development. MRS Bulletin, 41(3):231–240, 2016.

Chen Z, Beauvais M L, and Chapman K W. Pair distribution function analysis of discrete nanomaterials in PDFgui. J. Appl. Cryst., 56:328–337, 2023.

Chivers I and Sleightholme J. Introduction to Programming with Fortran. Springer International Publishing, Cham, 2018.

Chupas P J, Qiu, X, Hanson J C, Lee P L, Grey C P, and Billinge S J L. Rapid acquisition pair distribution function analysis (RA-PDF). J. Appl. Crystallogr., 36:1342–1347, 2003.

Chupas P J, Chapman K W, Kurtz C, Hanson J C, Lee P L, and Grey C P. A versatile sample-environment cell for non-ambient x-ray scattering experiments. J. Appl. Crystallogr., 41:822–824, 2008.

Clearfield A, and Smith G D. The crystallography and structure of α-zirconium bis(monohydrogen orthophosphate) monohydrate. Inorg. Chem., 8(3):431–436, 1969.

Cliffe M J, Dove M T, Drabold D A, and Goodwin A L. Structure determination of disordered materials from diffraction data. Phys. Rev. Lett., 104(12):125501, 2010.

Coelho A A. Topas and topas-academic: An optimization program integrating computer algebra and crystallographic objects written in C++. J. Appl. Crystallogr., 51(1):210–218, 2018.

Debye P. Dispersion of Røntgen rays. Ann. Phys. (Berl), 351:809–823, 1915.

Diebold U. The surface science of titanium dioxide. Surf. Sci. Rep., 48(5–8):53–229, 2003.

Dietl T and Ohno H. Dilute ferromagnetic semiconductors: Physics and spintronic structures. Rev. Mod. Phys., 86:187–251, 2014.

Dinnebier R E, and Billinge S J L. Powder Diffraction: Theory and Practice. RSC Publishing, Cambridge, 2008.

Dinnebier R E, Leineweber A, and Evans J S O. Rietveld refinement: Practical Powder Diffraction Pattern Analysis using TOPAS De Gruyeter. Berlin, 2019. Boston, MA

Dippel A-C, Roelsgaard M, Boettger U, Schneller T, Gutowski O, and Ruett U. Local atomic structure of thin and ultrathin films via rapid high-energy x-ray total scattering at grazing incidence. IUCrJ, 6(2):290–298, 2019.

Egami T and Billinge S J L. Underneath the Bragg Peaks: Structural Analysis of Complex Materials. Elsevier, Amsterdam, 2nd edition, 2012.

Farrow C L, Juhás P, Liu J, Bryndin D, Božin E S, Bloch J, Proffen Th, and Billinge S J L. PDFfit2 and PDFgui: Computer programs for studying nanostructure in crystals. J. Phys: Condens. Mat., 19:335219, 2007.

Farrow C L, Bediako D K, Surendranath Y, Nocera D G, and Billinge S J L. Intermediate-range structure of self-assembled cobalt-based oxygen evolving catalysts. J. Amer. Chem. Soc., 135:6403–6406, 2013.

Farrow C L, Shi C, Juhás P, Peng X, and Billinge S J L. Robust structure and morphology parameters for CdS nanoparticles by combining small angle X-ray scattering and atomic pair distribution function data in a complex modeling framework. J. Appl. Crystallogr., 47:561–565, 2014.

Frandsen B A, and Billinge S J L. Magnetic structure determination from the magnetic pair distribution function (mPDF): Ground state of MnO. Acta Crystallogr. A, 71(3):325–334, 2015.

Frandsen B A, Yang X, and Billinge S J L. Magnetic pair distribution function analysis of local magnetic correlations. Acta Crystallogr. A, 70(1):3–11, 2014.

Frandsen B A, Brunelli M, Page K, Uemura Y J, Staunton J B, and Billinge S J L. Verification of Anderson superexchange in MnO via magnetic pair distribution function analysis and *ab initio* theory. Phys. Rev. Lett., 116:197204, 2016.

Frandsen B A, Gong Z, Terban M W, Banerjee S, Chen B, Jin C, Feygenson M, Uemura Y J, and Billinge S J L. Local atomic and magnetic structure of dilute magnetic semiconductor $(Ba,K)(Zn,Mn)_2As_2$. Phys. Rev. B, 94:094102, 2016.

Frandsen B A, Taddei K M, Yi M, Frano A, Guguchia Z, Yu R, Si Q, Bugaris D E, Stadel R, Osborn R, Rosenkranz S, Chmaissem O, and Birgeneau R J. Local orthorhombicity in the magnetic C_4 phase of the hole-doped iron-arsenide superconductor $Sr_{1-x}Na_xFe_2As_2$. Phys. Rev. Lett., 119:187001, 2017.

Frandsen B A, Taddei K M, Bugaris D E, Stadel R, Yi M, Acharya A, Osborn R, Rosenkranz S, Chmaissem O, and Birgeneau R J. Widespread orthorhombic fluctuations in the $(Sr,Na)Fe_2As_2$ family of superconductors. Phys. Rev. B, 98:180505, 2018.

Funnell N P, Wang Q, Connor L, Tucker M G, O'Hare D, and Goodwin A L. Structural characterisation of a layered double hydroxide nanosheet. Nanoscale, 6:8032–8036, 2014.

Furukawa K. The radial distribution curves of liquids by diffraction methods. Rep. Prog. Phys., 25(1):395–440, 1962.

Gagin A, Allen A J, and Levin I. Combined fitting of small- and wide-angle x-ray total scattering data from nanoparticles: Benefits and issues. J. Appl. Crystallogr., 47(2):619–629, 2014.

Gamez L, Terban M W, Billinge S J L, and Martinez-Inesta M. Modelling and validation of particle size distributions of supported nanoparticles using the pair distribution function technique. J. Appl. Crystallogr., 50(3):741–748, 2017.

Gemmi M, Mugnaioli E, Gorelik T E, Kolb U, Palatinus L, Boullay P, Hovmøller S, and Abrahams J P. 3D electron diffraction: The nanocrystallography revolution. ACS Cent. Sci., 5(8):1315–1329, 2019.

Gereben O, Jovari P, Temleitner L, and Pusztai L. A new version of the RMC++ Reverse Monte Carlo program, aimed at investigating the structure of covalent glasses. J. Optoelectron. Adv. M., 9:3021–3027, 2007.

Gimenez E N, Chater P A, Crevatin G, Dennis G, Fairley A, Horswell I, Omar D, Spiers J, and Tartoni N. Arc-detector: Design of a CdTe photon-counting detector for the x-ray pair distribution function beamline at Diamond Light Source. *IEEE Nuclear Science Symposium and Medical Imaging Conference (NSS/MIC)*, 2019. Manchester, UK October 2019

Gong Z, and Billinge S J L. Atomic pair distribution functions (PDFs) from textured polycrystalline samples: Fundamentals. arXiv:1805.10342 [cond-mat], 2018.

Gorelik T E, Neder R, Terban M W, Lee Z, Mu X, Jung C, Jacob T, and Kaiser U. Towards quantitative treatment of electron pair distribution function. Acta Crystallogr. A, 75:532–549, 2019.

Granlund L, Billinge S J L, and Duxbury P M. Algorithm for systematic peak extraction from atomic pair distribution functions. Acta Crystallogr. A, 71(4):392–409, 2015.

Hammersley A P. FIT2D: a multi-purpose data reduction, analysis and visualization program. J. Appl. Crystallogr., 49(2):646–652, 2016.

Harouna-Mayer S Y, Tao S, Gong Z, v Zimmermann M, Koziej D, Dippel A-C, and Billinge S J L. Real-space texture and pole-figure analysis using the 3D pair distribution function on a platinum thin film. IUCrJ, 9(5):594–603, 2022.

Hinrichsen B, Dinnebier R E, and Jansen M. Two-dimensional diffraction using area detectors. In Dinnebier R E, and Billinge S J L, editors, Powder diffraction: Theory and practice, pages 414–438. RSC publishing, 2008. Cambridge

Hinrichsen B. Two-dimensional x-ray powder diffraction. Ph.D. Thesis, 2007. Max-Planck-Institut für Festkörperforschung Stuttgart

Hoque M M, Vergara S, Das P P, Ugarte D, Santiago U, Kumara C, Whetten R L, Dass A, and Ponce A. Structural analysis of ligand-protected smaller metallic nanocrystals by atomic pair distribution function under precession electron diffraction. J. Phys. Chem. C, 123:19894–19902, 2019.

Hwang J, Melgarejo Z H, Kalay Y E, Kalay I, Kramer M J, Stone D S, and Voyles P M. Nanoscale structure and structural relaxation in $zr_{50}cu_{45}al_5$ bulk metallic glass. Phys. Rev. Lett, 108:195505, 2012.

Jacques S D M, Di Michiel M, Kimber S A J, Yang X, Cernik R J, Beale A M, and Billinge S J L. Pair distribution function computed tomography. Nat. Commun., 4:2536, 2013.

Jensen K M Ø, Christensen M, Juhás P, Tyrsted, C, Bøjesen E D, Lock N, Billinge S J L, and Iversen B B. Revealing the mechanisms behind SnO_2 nanoparticle formation and growth during hydrothermal synthesis: An in situ total scattering study. J. Am. Chem. Soc., 134:6785–6792, 2012.

Jensen K M Ø, Blichfeld A B, Bauers S R, Wood S R, Dooryhée E, Johnson D C, Iversen B B, and Billinge S L J. Demonstration of thin film pair distribution function analysis (tfPDF) for the study of local structure in amorphous and crystalline thin films. IUCrJ, 2(5):481–489, 2015.

Jeong I-K, Proffen T, Mohiuddin-Jacobs F, and Billinge S J L. Measuring correlated atomic motion using x-ray diffraction. J. Phys. Chem. A, 103:921–924, 1999.

Jeong I-K, Thompson J, Turner A M P, and Billinge S J L. PDFgetX: A program for determining the atomic pair distribution function from X-ray powder diffraction data. J. Appl. Crystallogr., 34:536, 2001.

Juelsholt M, Christiansen T L, and Jensen K M Ø. Mechanisms for tungsten oxide nanoparticle formation in solvothermal synthesis: From polyoxometalates to crystalline materials. J. Phys. Chem. C, 123(8):5110–5119, 2019.

Juhás P, Cherba D M, Duxbury P M, Punch W F, and Billinge S J L. Ab initio determination of solid-state nanostructure. Nature, 440(7084):655–658, 2006.

Juhás P, Davis T, Farrow C L, and Billinge S J L. PDFGETX3: A rapid and highly automatable program for processing powder diffraction data into total scattering pair distribution functions. J. Appl. Crystallogr., 46:560–566, 2013.

Juhás P, Farrow C L, Yang X, Knox K R, and Billinge S J L. Complex modeling: A strategy and software program for combining multiple information sources to solve ill-posed structure and nanostructure inverse problems. Acta Crystallogr. A, 71(6):562–568, 2015.

Juhás P, Louwen J N, van Eijck L, Vogt E T C, and Billinge, S J L. *PDFgetN3*: atomic pair distribution functions from neutron powder diffraction data using *ad hoc* corrections. J. Appl. Crystallogr., 51(5):1492–1497, 2018.

Keen D A and Goodwin A L. The crystallography of correlated disorder. Nature, 521(7552):303–309, 2015.

Kern A. Instrumentation for laboratory x-ray scattering techniques. In Gilmore C, Kaduk, J, and Schenk, H et al., editors, International Tables of Crystallography, volume H, pages 26–50. International Union of Crystallography, 2019. Chester, UK

Kieffer J, Ashiotis G, Deschildre A, Nawaz Z, Wright J P, Karkoulis D, and Picca F E. The fast azimuthal integration python library: pyFAI. J. Appl. Crystallogr., 48:510–519, 2015.

Kieffer J, Valls V, Blanc N, and Hennig C. New tools for calibrating diffraction setups. J. Synchrotron Rad., 27:558–566, 2020.

Kovyakh A, Banerjee S, Liu C-H, Wright C J, Li Y C, Mallouk T E, Feiden-hans'l R, and Billinge S J L. Towards scanning nanostructure x-ray microscopy. arXiv:2110.01656 [cond-mat.mtrl-sci], 2021.

Krayzman V, Levin I, Woicik J C, Proffen T, Vanderah T A, and Tucker M G. A combined fit of total scattering and extended x-ray absorption fine structure data for local-structure determination in crystalline materials. J. Appl. Crystallogr., 42(5):867–877, 2009.

Krivoglaz M A. X-Ray and Neutron Diffraction in Nonideal Crystals. Springer Science & Business Media, 2012.

Lappas A, Antonaropoulos G, Brintakis K, Vasilakaki M, Trohidou K N, Iannotti V, Ausanio G, Kostopoulou A, Abeykoon M, Robinson I K, and Bozin E S. Vacancy-Driven Noncubic Local Structure and Magnetic Anisotropy Tailoring in FexO−Fe3−δO4 Nanocrystals. Phys. Rev. X, 9(4):041044, 2019.

Larsen A H, Jørgen Mortensen J, Blomqvist J, Castelli I E, Christensen R, Dułak M, Friis J, Groves M N, Hammer B, Hargus C, Hermes E D, Jennings P C, Bjerre Jensen P, Kermode J, Kitchin J R, Leonhard Kolsbjerg E, Kubal J, Kaas-bjerg K, Lysgaard S, Bergmann Maronsson J, Maxson T, Olsen T, Pastewka L, Peterson A, Rostgaard C, Schiøtz J, Schütt O, Strange M, Thygesen K S, Vegge T, Vilhelmsen L, Walter M, Zeng Z, and Jacobsen K W. The atomic simulation environment—a python library for working with atoms. J. Phys. Condens. Matter, 29(27):273002, 2017.

Lee J H, Aydiner C C, Almer J, Bernier J, Chapman K W, Chupas P J, Haeffner D, Kump K, Lee P L, Lienert U, Miceli A, and Vera G. Synchrotron applications of an amorphous silicon flat-panel detector. J. Synchrotron Rad., 15:477–488, 2008.

Lindahl Christiansen T, Cooper S R, and Jensen K M Ø. There's no place like real-space: Elucidating size-dependent atomic structure of nanomaterials using pair distribution function analysis. Nanoscale Adv., 2:2234–2254, 2020.

Lindahl Christiansen T, Emil T S, Kjær E T S, Kovyakh A, Røderen, M L, Høj M, Vosch T, and Jensen K M Ø. Structure analysis of supported disordered molyb-denum oxides using pair distribution function analysis and automated cluster modelling. J. Appl. Crystallogr., 53(1):148–158, 2020.

Liu C-H, Janke E M, Li R, Juhas P, Gang O, Talapin D V, and Billinge S J L. saspdfPDF: pPair dDistribution fFunction analysis of nanoparticle assemblies from small-angle-scattering data. J. Appl. Crystallogr., 53:699–709, 2020.

Liu C-H, Wright C J, Gu R, Bandi S, Wustrow A, Todd P K, O'Nolan D, Beauvais M L, Neilson J R, Chupas P J, Chapman K W, and Billinge S J L. Validation of non-negative matrix factorization for rapid assessment of large sets of atomic pair distribution function data. J. Appl. Crystallogr., 54(3): 768–775, 2021.

Macrae C F, Sovago I, Cottrell S J, Galek P T A, McCabe P, Pidcock E, Platings M, Shields G P, Stevens J S, Towler M, and Wood P A. *Mercury 4.0*: From visualization to analysis, design and prediction. J. Appl. Crystallogr., 53(1): 226–235, 2020.

Madsen I C, Scarlett N V Y, Kleebergb R, and Knorr K. Quantitative phase analysis. In Gilmore, C, Kaduk, J, and Schenk, H., editors, International Tables of Crystallography, volume H, pages 344–373. International Union of Crystallography, Chester, UK, 2019.

Marlton F, Ivashko O, Zimmerman M, Gutowski O, Dippel A-C, and Jørgensen M R V. A simple correction for the parallax effect in x-ray pair distribution function measurements. J. Appl. Crystallogr., 52:1072–1076, 2019.

Masadeh A S, Božin E S, Farrow C L, Paglia G, Juhás P, Karkamkar A, Kanatzidis M G, and Billinge S J L. Quantitative size-dependent structure and strain determination of CdSe nanoparticles using atomic pair distribution function analysis. Phys. Rev. B, 76:115413, 2007.

McGreevy R L, and Zetterstrom P. To rmc or not to rmc? The use of reverse Monte Carlo modelling. Curr. Opin. Solid State Mater. Sci., 7(1):41–47, 2003.

Mcgreevy R L, Howe M A, Keen D A, and Clausen K N. Reverse Monte-Carlo (RMC) simulation - modeling structural disorder in crystals, glasses and liquids from diffraction data. IOP Conference Series, 107:165–184, 1990.

McGreevy R L. Reverse Monte Carlo modelling. J. Phys.: Condens, Matter, 13(46):R877–R913, 2001.

Metz P C, Koch R, and Misture S T. Differential evolution and Markov chain Monte Carlo analyses of layer disorder in nanosheet ensembles using total scattering. J. Appl. Crystallogr., 51:1437–1444, 2018.

Momma K and Izumi F. *VESTA3* for three-dimensional visualization of crystal, volumetric and morphology data. J. Appl. Crystallogr., 44(6):1272–1276, 2011.

Mou Q, Benmore C J, Weber W S, and Yarger J L. Insights into the hierarchical structure of spider dragline silk fibers: Evidence for fractal clustering of β-sheet nano-crystallites. arXiv:1507.04321 [cond-mat.mtrl-sci], 2015.

Munro D H, and Dubois P F. Using the Yorick Interpreted Language. Computers in Physics, 9(6):609–615, 1995.

Nakamura N, Terban M W, Billinge S J L, and Reeja Jayan B. Unlocking the structure of mixed amorphous-crystalline ceramic oxide films synthesized under low temperature electromagnetic excitation. J. Mater. Chem. A, 5(35):18434–18441, 2017.

Nakamura N, Su L, Wang H, Bernstein N, Jha S K, Culbertson E, Wang H, Billinge S J L, Hellberg C S, and Reeja-Jayan B. Linking far-from-equilibrium defect structures in ceramics to electromagnetic driving forces. J. Mater. Chem. A., 9(13):8425–8434, 2021.

Neder R B, and Proffen T. Diffuse Scattering and Defect Structure Simulations: A Cook Book using the Program DISCUS. Oxford University Press, Oxford, 2008.

Nield V M, and Keen D A. Diffuse Neutron Scattering from Crystalline Materials. Clarendon Press, 2001.

Olds D, Peterson P F, Crawford M K, Neilson J R, Wang H-W, Whitfield P S, and Page K. Combinatorial appraisal of transition states for *in situ* pair distribution function analysis. J. Appl. Crystallogr., 50(6):1744–1753, 2017.

O'Nolan D, Huang G, Kamm G E, Grenier A, Liu C-H, Todd P K, Wustrow A, Tran G T, Montiel D, Neilson J R, Billinge S J L, Chupas P J, Thornton K S, and Chapman K W. A thermal-gradient approach to variable-temperature measurements resolved in space. J. Appl. Cryst., 53:662–670, 2020.

Palmer D C. Visualization and analysis of crystal structures using CrystalMaker software. Z. Kristallogr., 230:559–572, 2015.

Peterson P F, Gutmann M, Proffen T, and Billinge S J L. PDFgetN: A user-friendly program to extract the total scattering structure function and the pair distribution function from neutron powder diffraction data. J. Appl. Crystallogr., 33(4):1192–1192, 2000.

Petkov V, DiFrancesco R G, Billinge S J L, Acharya M, and Foley H C. Local structure of nanoporous carbons. Philos. Mag. B, 79:1519, 1999.

Petkov V. RAD, a program for analysis of x-ray diffraction data from amorphous materials for personal computers. J. Appl. Crystallogr., 22(4):387–389, 1989.

Prescher C, and Prakapenka V B. Dioptas: A program for reduction of two-dimensional x-ray diffraction data and data exploration. High Press. Res., 35(3):223–230, 2015.

Prill D, Juhás P, Schmidt M U, and Billinge S J L. Modeling pair distribution functions (PDF) of organic compounds: describing both intra- and intermolecular correlation functions in calculated PDFs. J. Appl. Crystallogr., 48(1):171–178, 2015.

Prill D, Juhás P, Billinge S J L, and Schmidt M U. Towards solution and refinement of organic crystal structures by fitting to the atomic pair distribution function. Acta Crystallogr. A, 72(1):62–72, 2016.

Proffen T and Billinge S J L. PDFFIT, a program for full profile structural refinement of the atomic pair distribution function. J. Appl. Crystallogr., 32:572–575, 1999.

Proffen T, and Neder R B. *DISCUS*: A program for diffuse scattering and defect-structure simulation. J. Appl. Crystallogr., 30(2):171–175, 1997.

Qiu X, Thompson J W, and Billinge S J L. PDFgetX2: A GUI driven program to obtain the pair distribution function from X-ray powder diffraction data. J. Appl. Crystallogr., 37:678, 2004.

Radaelli P G, Horibe Y, Gutmann M J, Ishibashi H, Chen C H, Ibberson R M, Koyama Y, Hor Y S, Kiryukhin V, and Cheong S W. Formation of isomorphic Ir^{3+} and Ir^{4+} octamers and spin dimerization in the spinel $CuIr_2S_4$. Nature, 416:155–158, 2002.

Rakita Y, Hart J L, Das P P, Shahrezaei S, Foley D L, Mathaudhu S N, Nicolopoulos S, Taheri M L, and Billinge S J L. Mapping structural heterogeneity at the nanoscale with scanning nano-structure electron microscopy (SNEM). Acta Mater., 242:118426, 2022.

Reeja-Jayan B, Harrison K L, Yang K, Wang, C-L, Yilmaz A E, and Manthiram A. Microwave-assisted low-temperature growth of thin films in solution. Sci. Rep., 2:1003, 2012.

Roelsgaard M, Dippel A-H, Borup K A, Nielsen I G, Broge N L N, Røh J T, Gutowski O, and Iversen B B. Time-resolved grazing-incidence pair distribution functions during deposition by radio-frequency magnetron sputtering. IUCrJ, 6:299–304, 2019.

Schlomberg H, Kröger J, Savasci G, Terban M W, Bette S, Moudrakovski I, Duppel V, Podjaski F, Siegel R, Senker J, Dinnebier R E, Ochsenfeld C, and Lotsch B V. Structural insights into poly(heptazine imides): A light-storing carbon nitride material for dark photocatalysis. Chem. Mater., 31(18):7478–7486, 2019.

Schökel A, Etter M, Berghaüser A, Horst A, Lindackers D, Whittle T A, Schmid S, Acosta M, Knapp M, Ehrenberg H, and Hinterstein M. Multi-analyser detector (MAD) for high-resolution and high-energy powder x-ray diffraction. J. Synchrotron Rad., 28:146–157, 2021.

Sclafani A and Herrmann J M. Comparison of the photoelectronic and photocatalytic activities of various anatase and rutile forms of titania in pure liquid organic phases and in aqueous solutions. J. Phys. Chem., 100(32):13655–13661, 1996.

Shannon R D. Revised effective ionic radii and systematic studies of interatomic distances in halides and chalcogenides. Acta Crystallogr. A., 32:751–767, 1976.

Shull C G, Strauser W A, and Wollan E O. Neutron diffraction by paramagnetic and antiferromagnetic substances. Acta Crystallogr. A., 83(2):333–345, 1951.

Silbernagel R, Martin C H, and Clearfield A. Zirconium(iv) phosphonate–phosphates as efficient ion-exchange materials. Inorg. Chem., 55(4):1651–1656, 2016.

Silbernagel R, Shehee T C, Martin C H, Hobbs D T, and Clearfield A. Zr/Sn(iv) phosphonates as radiolytically stable ion-exchange materials. Chem. Mater., 28(7):2254–2259, 2016.

Skinner L B, Benmore C J, and Parise J B. Area detector corrections for high quality synchrotron x-ray structure factor measurements. Nucl. Instrum. Methods Phys. Res., 662:61–70, 2012.

Skinner L B, Benmore C J, and Parise J B. Comment on 'molecular arrangement in water: Random but not quite'. J. Phys,: Condens. Matter, 24:155102, 2012.

Soper A K. Empirical potential Monte Carlo simulation of fluid structure. Chem. Phys., 202:295–306, 1996.

Soper A K. GudrunN and GudrunX: Programs for correcting raw neutron and x-ray total scattering data to differential cross section. 2012.

Stroustrup B. A history of C++: 1979–1991. In Bergin, T J, and Gibson, R, editors, History of Programming Languages—II, pages 699–769. Association for Computing Machinery, New York, NY, USA, January 1996.

Taddei K M, Allred J M, Bugaris D E, Lapidus S, Krogstad M J, Stadel R, Claus H, Chung D Y, Kanatzidis M G, Rosenkranz S, Osborn R, and Chmaissem O. Detailed magnetic and structural analysis mapping a robust magnetic C_4 dome in $Sr_{1-x}Na_xFe_2As_2$. Phys. Rev. B, 93:134510, 2016.

Terban M W, and Billinge S J L. Structural analysis of molecular materials using the pair distribution function. Chemical Reviews, 122(1):1208–1272, 2022.

Terban M W, Johnson M, DiMichiel M, and Billinge S J L. Detection and characterization of nanoparticles in suspension at low concentrations using the x-ray total scattering pair distribution function technique. Nanoscale, 7:5480–5487, 2015.

Terban M W, Shi C, Silbernagel R, Clearfield A, and Billinge S J L. Local environment of terbium(iii) ions in layered nanocrystalline zirconium(iv) phosphonate–phosphate ion exchange materials. Inorg. Chem., 56(15):8837–8846, 2017.

Thatcher Z, Liu C-H, Yang L, McBride B C, Tran G T, Wustrow A, Karlsen M A, Neilson J R, Ravnsbæk D B, and Billinge S J L. nmfMapping: A cloud-based web application for non-negative matrix factorization of powder diffraction and pair distribution function datasets. Acta Crystallogr. A., 78(3) 2022.

Thi Y N, Rademann K, and Emmerling F. Direct evidence of polyamorphism in paracetamol. CrystEngComm., 17(47):9029–9036, 2015.

Toby B H and Billinge S J L. Determination of standard uncertainties in fits to pair distribution functions. Acta Crystallogr. A, 60:315–317, 2004.

Toby B H and Egami T. Accuracy of pair distribution function analysis applied to crystalline and noncrystalline materials. Acta Crystallogr. A, 48(3):336–46, 1992.

Toby B H and Von Dreele R B. *GSAS-II*: the genesis of a modern open-source all purpose crystallography software package. J. Applied Crystallogr., 46(2): 544–549, 2013.

Toby B. Rietveld refinemnet. In Gilmore, C, Kaduk, J, and Schenk, H, editors, International Tables of Crystallography, volume H, pages 465–472. International Union of Crystallography, Chester, UK, 2019.

Treacy M M, Newsam J M, and Deem M W. A general recursion method for calculating diffracted intensities from crystals containing planar faults. Proc. R. Soc. Lond. A, 433:499–520, 1991.

Tucker M G, Keen D A, Dove M T, Goodwin A L, and Hui Q. RMCProfile: Reverse Monte Carlo for polycrystalline materials. J. Phys,: Condens. Matter, 19:335218, 2007.

Van Rossum G and Drake F L. Python 3 Reference Manual. CreateSpace, Scotts Valley, CA, 2009.

Wagner C N J. Direct methods for the determination of atomic-scale structure of amorphous solids (x-ray, electron and neutron scattering). J. Non-Cryst. Solids, 31:1, 1978.

Warren B E. X-ray diffraction in random layer lattices. Phys. Rev., 59(9):693–698, 1941.

Weber T and Simonov A. The three-dimensional pair distribution function analysis of disordered single crystals: Basic concepts. Z. Kristallogr. Cryst. Mater., 227(5):238–247, 2012.

Weber J K R, Rey C A, Neuefeind J, and Benmore C J. Acoustic levitator for structure measurements on low temperature liquid droplets. Rev. Sci. Instrum., 80:083904, 2009.

Weber R J, Benmore C J, Tumber S K, Tailor A N, Rey C A, Taylor L S, and Byrn S R. Acoustic levitation: Recent developments and emerging opportunities in biomaterials research. Eur. Biophys. J., 41:397–403, 2012.

Welberry T R. Diffuse. X-Ray Scattering and Models of Disorder. Number 16 in International Union of Crystallography Monographs on Crystallography. Oxford University Press, Oxford, 2010.

Wright C J and Zhou X-D. Computer-assisted area detector masking. J. Synchrotron Rad., 24:506–508, 2017.

Yang X, Masadeh A S, McBride J R, Božin E S, Rosenthal S J, and Billinge S J L. Confirmation of disordered structure of ultrasmall CdSe nanoparticles from x-ray atomic pair distribution function analysis. Phys. Chem. Chem. Phys., 15(22):8480–8486, 2013.

Yang X, Juhás P, and Billinge S J L. On the estimation of statistical uncertainties on powder diffraction and small-angle scattering data from two-dimensional x-ray detectors. J. Appl. Crystallogr., 47(4):1273–1283, 2014.

Yang X, Juhás P, Farrow C, and Billinge S J L. xPDFsuite: An end-to-end software solution for high throughput pair distribution function transformation, visualization and analysis. arXiv:1402.3163 [cond-mat.mtrl-sci], 2015.

Yang L, Juhás P, Terban M W, Tucker M G, and Billinge S J L. Structure-mining: Screening structure models by automated fitting to the atomic pair distribution function over large numbers of models. Acta Crystallogr. A., 76(3):395–409, 2020.

Yang L, Culbertson E A, Thomas N K, Vuong H T, Kjær E T S, Jensen K M Ø, Tucker M G, and Billinge S J L. A cloud platform for atomic pair distribution function analysis: PDFitc. Acta Crystallogr. A., 77(1):2–6, 2021.

Young C A and Goodwin A L. Applications of pair distribution function methods to contemporary problems in materials chemistry. J. Mater. Chem., 21:6464–6476, 2011.

Zhao K, Deng Z, Wang X C, Han W, Zhu J L, Li X, Liu Q Q, Yu R C, Goko T, Frandsen B, Frandsen Liu, L, Ning F, Uemura Y J, Dabkowska H, Luke G M, Luetkens H, Morenzoni E, Dunsiger S R, Senyshyn A, Boeni P, and Jin C Q. New diluted ferromagnetic semiconductor with Curie temperature up to 180 K and isostructural to the '122' iron-based superconductors. Nat. Commun., 4:1442, 2013.

Zhao X-G, Malyi O I, Billinge S J L, and Zunger A. Intrinsic local symmetry breaking in nominally cubic paraelectric $BaTiO_3$. Phys. Rev. B., 105(22):224108, 2022.

Index